Simple Model Railway Layouts

Simple Model Railway Layouts

T.J.Booth

Patrick Stephens
Wellingborough, Northamptonshire

© T. J. Booth 1987

All rights reserved. No part of this publication may be reproduced, stored in a retrieval system or transmitted, in any form or by any means, electronic, mechanical, photocopying, recording or otherwise, without prior permission in writing from Patrick Stephens Limited.

First published in 1987

British Library Cataloguing in Publication Data

Booth, T.J.
    Simple model railway layouts.
    1. Railroads — Models
    I. Title
    625.1'9   TF197

ISBN 0-85059-878-8

Patrick Stephens Limited is part of the Thorsons Publishing Group

Printed and bound in Great Britain

10 9 8 7 6 5 4 3 2 1

Contents

Acknowledgements	6
Introduction	7
CHAPTER 1 Standard gauge light railway	9
CHAPTER 2 OO gauge branch line terminus	23
CHAPTER 3 OO gauge 'modern scene' layout	39
CHAPTER 4 N gauge main line	56
CHAPTER 5 An American 'shortline'	72
CHAPTER 6 009 narrow gauge quarry layout	89
CHAPTER 7 French branch line terminus	102
CHAPTER 8 0 gauge light railway	120
CHAPTER 9 Railway Preservation Centre	136
CHAPTER 10 N gauge North American layout	148
Index	157

Acknowledgements

I am grateful to the following companies and individuals for their invaluable help and assistance in providing photographs and information.

Ratio Plastic Models Ltd
Richard Kohnstam Ltd
Peco
N. & K.C. Keyser Ltd
BTA Hobbies

Ian Allan
Ron Cadman (modelmaker)
Scale Models and Kits, Bolton
Nick Wood

Special thanks are due to Mary Fletcher, Bill Morton, Terry Livesey and Louise Bleasdale for their help with the photographs and drawings respectively.

Introduction

With the abundance of excellent models, equipment, books, magazines and plans available, it is often difficult to know where to start in such a wide and diverse hobby as railway modelling. There are many excellent books covering in detail individual aspects such as locomotive construction, scenic modelling and electronics, but the intention of this book is to present ideas for small, fairly simple layouts which should not prove too expensive to construct and should be capable of finding a place in the modern home, two very important and often overlooked considerations!

With each of the layouts there is a guide as to how it may be constructed in the hope that it will encourage would-be layout builders to have a go. These are, however, intended purely to present ideas and suggestions and not to deliver either a blow-by-blow account or a 'tablet of stone' which must be slavishly adhered to. The methods suggested have all been well tried and are fairly common; in most cases there are also alternatives, and all but the most inexperienced modellers will no doubt have developed their own favourite approaches and methods.

Each chapter in the book is self-contained and provides an idea and a plan for a layout and suggests ways in which it might be constructed and the locomotives and rolling-stock which might be appropriate. However, it is quite possible that a suggestion or idea in one chapter could equally well be used in another.

No apology is made for dealing with baseboard construction in some detail, as the baseboard is in every sense the foundation upon which all else depends. Skimping on baseboards and careless track laying are two common causes of poor running on model railway layouts, and any short cuts and economies that are tried in these areas will undoubtedly prove false and probably costly to remedy.

The layout plans included are not revolutionary but are based on ideas and concepts which have been well used in the hobby over a long period. They are of necessity a compromise between reality and the practicality of constructing a small working model railway, but similar layouts have either been built by the author or he has been associated with them, and, despite their modest size, they can all become satisfying models.

Because of the scale, the layout plans cannot be entirely accurate; they are suggestions only, and because of their small size great care needs to be taken to ensure that adequate clearances and maximum lengths of sidings, run-round loops and the like are obtained. They can be accommodated in the spaces shown but would, of course, benefit from a larger area. The builder can therefore adapt the plans to suit his own preferences and the space available; they could be extended or perhaps later accommodated into a bigger system.

Because they are small and relatively cheap to construct, these layouts provide an ideal 'learning' exercise by which to practise and improve the various skills and techniques used in the hobby and can be scrapped and another layout built without too much concern. Small layouts can also be detailed to a far higher and more uniform standard than larger systems and have a much greater chance of being completed; a large layout can often prove too much for the individual modeller. Above all, small layouts are an ideal testing-ground for the techniques we will use in the 'ideal' layout of which we all dream!

At the end of each chapter is a list of some of the many sources of further information available which relate to the ideas dealt with in that chapter. These are not exhaustive and include books which may be out of print, but your local library should be able to get copies for you and if you would like to have a copy of your own you may be able to trace one through a second-hand book dealer (there is now a good trade in second-hand railway books). Similarly, each chapter has specific references to a variety of products, and whilst every effort has been made to ensure that they are available at the time of publication, no guarantee can be made. Usually, however, good second-hand examples of many of the items of rolling-stock and locomotives can be found and for those particularly concerned with the cost of building a model railway layout, great savings can be made by buying second-hand from a reputable dealer.

Further information, help and advice is readily available from books and magazines and by asking questions at model railway exhibitions. If you have a local model railway club, the members will provide a ready source of help and advice. Equally, specialist societies such as the Gauge O Guild, EM Society and SNCF Society are there to help promote their particular interests and can be a useful source of help and expertise if your interests develop in those directions. Lastly, do not be afraid to ask the manufacturers themselves who in most cases are only too willing, for the consideration of a stamped addressed envelope, to give information on their products.

If this book provides some help and encouragement to would-be layout builders, and a few more model railway layouts are built, then it will have served its purpose.

CHAPTER 1
Standard gauge light railway

Away from the glamour of Euston, Kings Cross, Waterloo and the rest of Britain's main line railway system, there existed an almost twilight world of small railways running on shoestring budgets, begging, borrowing and, I dare say, stealing to eke out a paltry existence. Often serving remote rural populations, their origins and raison d'être had long since been lost. These were Britain's light railways, independent railways usually built to standard gauge, but with notable exceptions such as the 15in gauge of the Romney, Hythe and Dymchurch.

They were very much the poor relations of our country's railways, built and run on the cheap and often existing for a purely local and specific purpose. They seldom had any delusions of grandeur but they did provide a local service and fulfilled a local need, filling in the gaps in the main line railway system. Perhaps the best-known light railway, the Kent and East Sussex, continues to run at least in part to the present day.

The rules for running a main line railway did not apply to the systems operating under a Light Railway Order, so they provide an excellent subject for the railway modeller, especially one with a limited space for his layout. In many respects, the 'light railway' model is a licence to let the imagination run riot. Most light railways, certainly before nationalization in 1948, relied on second-hand locomotives and rolling-stock to run their services; some railways 'borrowed' suitable locomotives from the main line companies, often not being able to afford to repair or replace their own aged locomotives and stock.

Light railways were built to a lower standard than their main line counterparts; bridges and other structures were limited to the bare essentials, both in quantity and strength.

Therefore, the light railway, with its reliance on old stock and light construction, means that a realistic and representative model can be made in a small space. Also, older coaches and locomotives tend to be smaller than their more modern counterparts, thus the space required for platforms, run-round loops and so on is less than for normal railways. Short trains, often composed of a single bogie coach or two four-wheelers, and mixed trains were the order of the day.

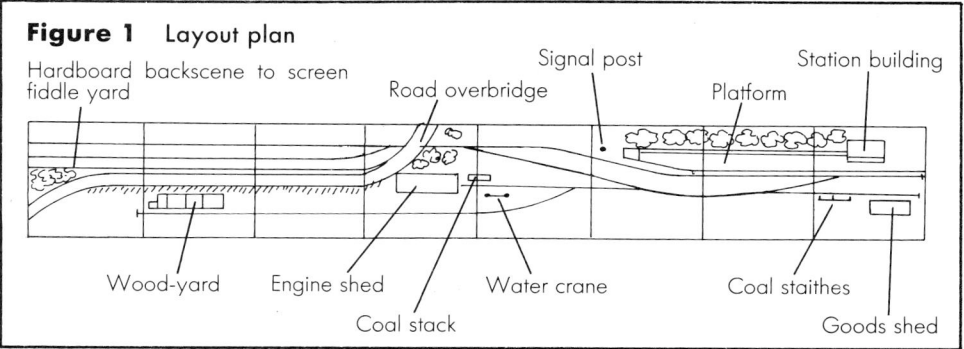

Figure 1 Layout plan

The plan

The layout plan (Figure 1) is for a terminus on a British light railway. It is a fairly basic facility, occupying a space of some 8ft × 1ft. As with all the layouts in this book, a little more available space would allow for the lengthening of the run-round facilities and thus longer trains, while an inch or two on the width would increase the scope for scenic development. However, there is sufficient interest in building and running this layout to keep the average modeller occupied for a while, and it could, should the opportunity arise at a later date, be either extended or perhaps incorporated into a larger layout.

Baseboard construction

It is desirable to make small layouts as portable as possible, especially if they are to be erected only when in use, being stored in the meanwhile. It is also possible that if the layout is built and finished to a high standard its appearance at a model railway exhibition may be a possibility. There is a limit to the size of baseboards which can be regarded as 'portable', particularly for the individual enthusiast, and thus it is preferable to make this model, even though it is only 8ft long, in more than one section. Minimising rail and baseboard joints and thus electrical joints is a help in maintaining optimum reliability, so this layout is constructed on just two equal baseboards each 4ft × 1ft. The baseboards are bolted together as shown in Figure 2 for storage and carriage.

Weight is an important consideration, and 12mm plywood has been chosen for the framework, with Sundeala type board for the upper surface. Sundeala is a type of composition board which is fairly strong, cuts easily and is very easy to stick track pins into. The plywood framework is light and strong, particularly if the centres of the cross-pieces are removed with a jigsaw. Above all, it is not likely to warp or twist like the more traditional softwood frame. Plywood and Sundeala are both quite expensive if bought in full sheets, but this layout calls for much less than a full sheet and part sheets can be bought at most DIY shops at a reasonable cost.

Squareness and accuracy of cut for the baseboard materials is essential, not merely for appearance but as an essential prerequisite for the satisfactory working of the layout. It is a false

Standard gauge light railway — 11

Figure 2 Storage and carriage of baseboard

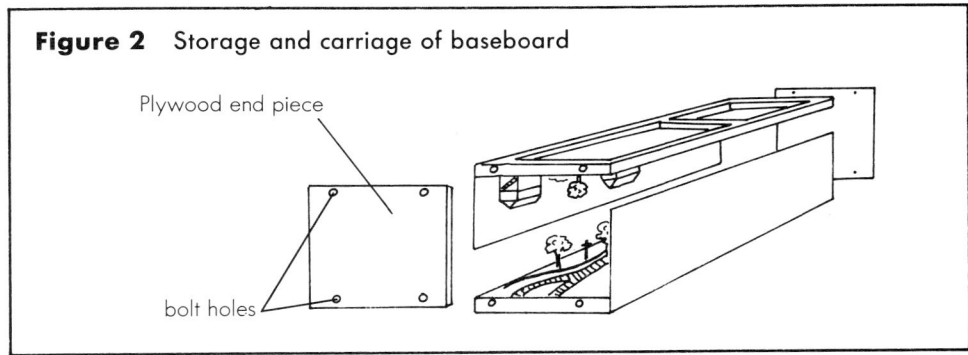

economy to skimp on baseboards as they are quite literally the foundation on which a layout is built. Therefore, unless you have the confidence and the appropriate sharp tools, have the baseboard materials cut to size where bought. Most DIY stores will do this for a nominal charge which will not significantly add to the cost of this layout (at least one national chain of DIY stores not only sells part sheets of plywood and a Sundeala-type material but also offers a free cutting service for timber purchased at the store).

Figure 3 Baseboard construction

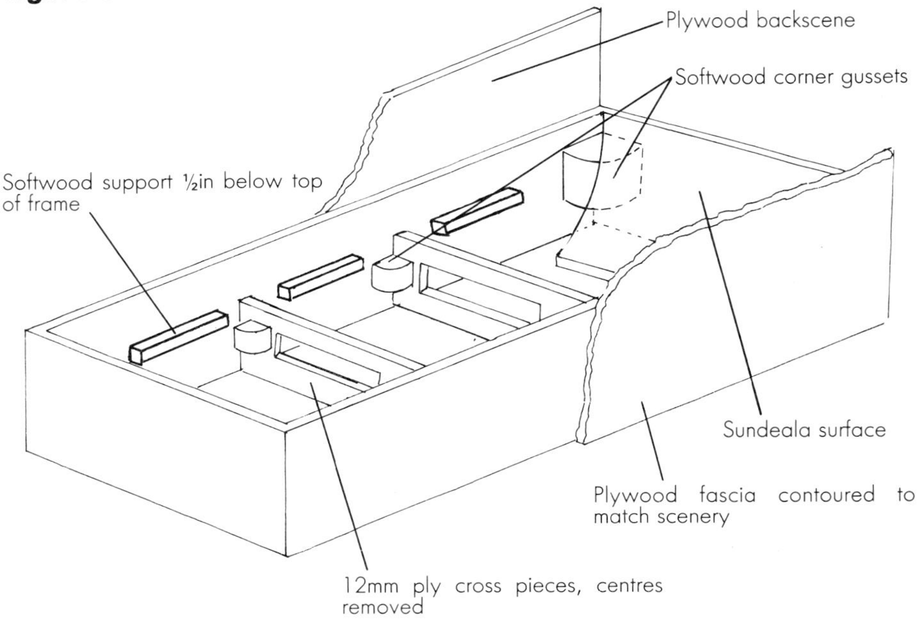

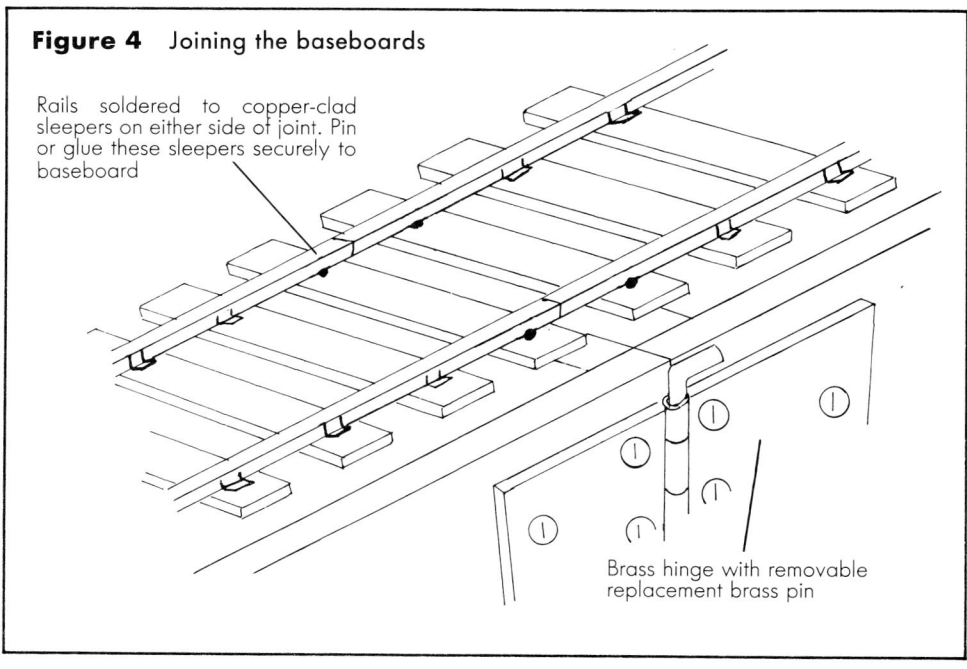

Figure 4 Joining the baseboards

Rails soldered to copper-clad sleepers on either side of joint. Pin or glue these sleepers securely to baseboard

Brass hinge with removable replacement brass pin

Figure 3 illustrates the baseboard construction. The method of joining the baseboards is simple and has the advantages of being self-contained, neat and uncomplicated (Figure 4). It consists of hinges screwed to the side members of the baseboards across the joints. Brass hinges are preferable to steel, not just for their appearance but because they will not rust. The head of the hinge-pin is filed off and the pin is removed to be replaced by a piece of brass rod somewhat longer than the original pin and bent to form a handle. This enables the boards to be separated. It is essential that the replacement pin is a tight fit to maintain baseboard alignment.

Trackwork

Light railway trackwork was usually of a smaller section than main line rail, and also often had ash or local stone as ballast. There are a number of flexi track systems on the market which represent the older-type 'bullhead' rail which is still to be found on the secondary lines of British Rail. Any of these would be suitable for this layout, but to date there are no matching ready-made points available. However, all is not lost as SMP Scaleway produce a wide selection of point kits to match their trackwork, including a simple-to-construct plastic-sleepered 3ft radius turnout, ideal for this layout. As an alternative to building your own pointwork, the model railway magazines often carry advertisements for ready-assembled pointwork to match these finer-scale track systems. Obviously more expensive than doing it yourself, the basic 3ft radius turnouts required for this layout are still very reasonably

Careful planning is essential, particularly with small layouts. Use cheap decorator's lining-paper to draw out the layout full size, marking the baseboard joints and framing; problems with locating point motors and other items beneath the baseboard can thus be avoided. Positioning of trackwork to maximize the length of run-round loops and sidings can also be worked out, and you don't need to have completed buildings and structures to give an idea of the overall effect — this can be achieved using boxes and odd items of stock to work out such details as siding lengths.

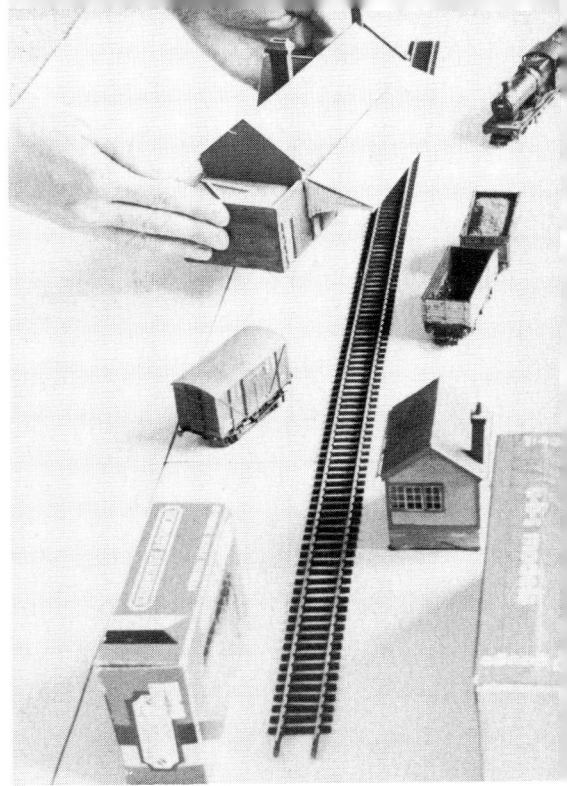

priced, and can be obtained wired ready for installation.

As with all small layouts, great care is needed in setting out the track plan on the baseboard to allow for maximum clearances and maximum run-round loop length and siding capacity. The track plan will help give the approximate position but trial-and-error placing of point-work and a length or two of track with some items of rolling-stock and, if no buildings are available, books and boxes will help you get the feel of the layout before anything is fixed down and will hopefully save problems and frustration later. If it is intended to place point motors beneath the baseboard surface, be careful not to place the tie bar of the pointwork over the baseboard frame, and avoid placing the points across the joint between the two baseboard sections.

Once the final position of the trackwork has been established, it should be marked on the baseboard surface, as should the position of any other fixtures or buildings decided upon at this stage. However, the important thing is the position of the trackwork. For this layout, laying it directly on the baseboard surface will help to give the required appearance of light section trackwork. Laying the track and ballasting are carried out as part of the same process, in the following stages:

1. Carefully mark where the track is to be laid on the baseboard.
2. Cut the trackwork and make any alterations such as slicing off moulded chairs to allow for the use of fish-plates.
3. Paint the area where the track is to go with PVA woodworking glue,

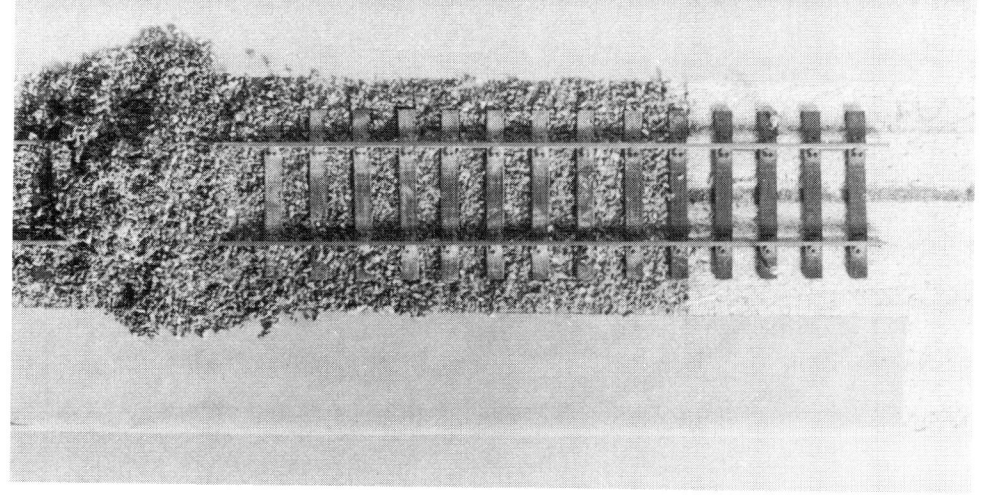

The finer trackwork systems available for 00 gauge such as the SMP illustrated here have a far better appearance than the Code 100 flat-bottom rail types, and will take all modern rolling-stock and locomotives. The photograph shows the easy process of laying the track directly on to the baseboard, as required for this layout to represent a shallow-ballasted light railway. The area the track is to occupy is coated with PVA woodworking glue, then the track is laid to the centre of the marked line and lightly pinned down. The ballast is spread over the track and the surplus brushed away 24 hours later. Tape can be used to mask the adjoining ground to ensure that the ballast goes only where required.

ensuring an even coat, but leave a gap where the point tie-bars are to go.

4 Place the track in position, placing fishplates where appropriate and making final adjustments to ensure good alignment of rails. Lightly pin down, then check the alignment again.
5 Add the ballast over the track and glued area.
6 Leave overnight to dry thoroughly.

Do not try to do the whole layout in one go, but do it section by section, taking time and care. Carelessly-laid or badly-aligned trackwork will prevent good running.

Fine granulated cork ballast is preferable for this layout, where the track is laid directly on the baseboard surface, as it tends to give quieter running than granite chips. Ensure that the ballast used is the finest available. Use fine ash or a similar mix of scenic material rather than ballast in the goods yard area, and coal dust around the engine shed. Add some 'grass' scenic scatter to the track ballast or ash mix at the ends of the sidings to give the impression of weeds.

Once the glue is dry, remove any surplus ballast material from the baseboard, clean the trackwork and ensure that the point blades move freely. Roll some stock over the layout to check running and make any necessary adjustments. File the rail joints lightly if necessary to smooth the running.

Wiring

The wiring for this layout is very simple and for convenience can be divided

Standard gauge light railway 15

into two aspects, powering the track and operating the turnouts. At its simplest, utilising the self-isolating properties of the kit points referred to earlier, only three feeds are required to the track to operate the layout via simple on – off switches from the controller (Figure 5).

There are a variety of point motors on the market which will operate the points from beneath the baseboard. Follow the manufacturers instructions supplied with the individual products. The more expensive units have substantial electrical switching facilities which can be used for a variety of purposes. Most importantly they can be used to switch the electrical polarity of points when they are operated and can either be used instead of the crude but effective arrangement incorporated in the points or as an additional 'belt and braces' back-up. Whatever make of motor is used, the wiring should be linked to the control panel via a suitable passing contact switch or a simple 'stud and probe' to provide the

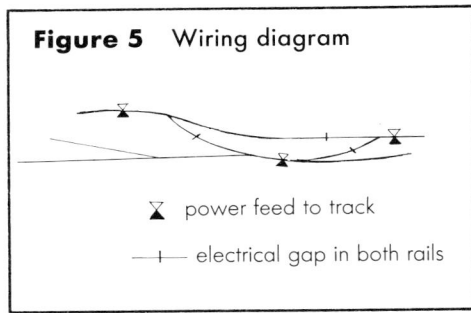

Figure 5 Wiring diagram

X power feed to track

—|— electrical gap in both rails

required turnout operation.

The wiring suggested is deliberately simple and based on the principle that, provided there is sufficient power to cover operating needs, the less there is, the less can go wrong. No doubt more complex and sophisticated wiring could be provided, but that is outside the scope of this book, the aim of which is to keep things simple.

The wiring from the layout should be harnessed neatly together and fastened to the underframe, ending in a multi-pin socket on the baseboard edge. The wiring is then carried by multi-pin plugs and harnessed to the

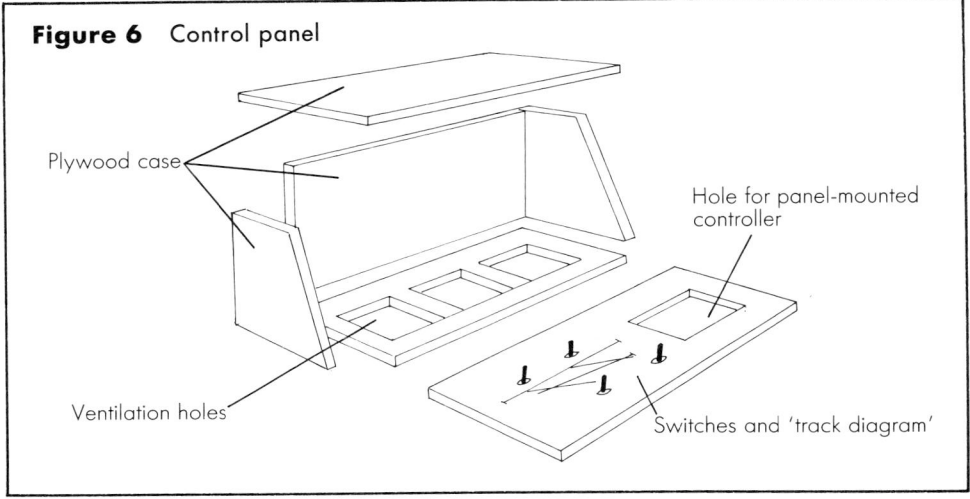

Figure 6 Control panel

control panel (Figure 6). In anticipation of possible future expansion, it is a good idea to use multi-pin plugs and sockets of greater capacity than immediate needs demand. The cable from the control panel should be long enough to reach under the baseboard so that if the layout is ever exhibited it can have the control panel hooked on the back but still use the existing plug and socket arrangement.

Scenery

There is not a lot of room for scenery in this layout beyond the immediate environs of the railway itself. However, with careful consideration, the use of simple techniques and the many excellent scenic details available from the trade, it is quite a straightforward job to give the impression of a run-down rural backwater at whatever period is chosen.

Before plunging in with the scenic scatter and paint, some consideration should be given to the period and the area it is wished to model. Some light railways had particular features and styles all their own, perhaps the most well documented being those of Colonel Stephens' empire, and the benefit of modelling a light railway is that you can invent your own 'house style' as you go along. Some suggestions for structures, including the station facilities, goods warehouse and engine shed are given here.

The *station platform* is a simple construction using balsa and cardboard, and Figure 7 illustrates the construction. Period is not only important when considering locomotives and rolling-stock but also when adding scenic details. Choose them carefully and they will give a very nice period feel to a scene. For example, an out-of-

Northiam station on the preserved section of the Kent and East Sussex Railway. This type of building was common on a number of lines in the south-east engineered by Col H. F. Stephens, and would be suitable for this layout.

Standard gauge light railway 17

The Peco 'Manyways' station building makes a quite realistic model and would be ideal for this type of layout. (Peco Ltd)

period road vehicle or horse-drawn carriage, however nice a model in its own right, would spoil the effect of a 1960s layout. There are many excellent detailed figures now available and their costume clearly highlights their period and in some instances also their trade.

The only other major engineering work on the layout is the *overbridge* behind the wood-yard. Figure 8 shows

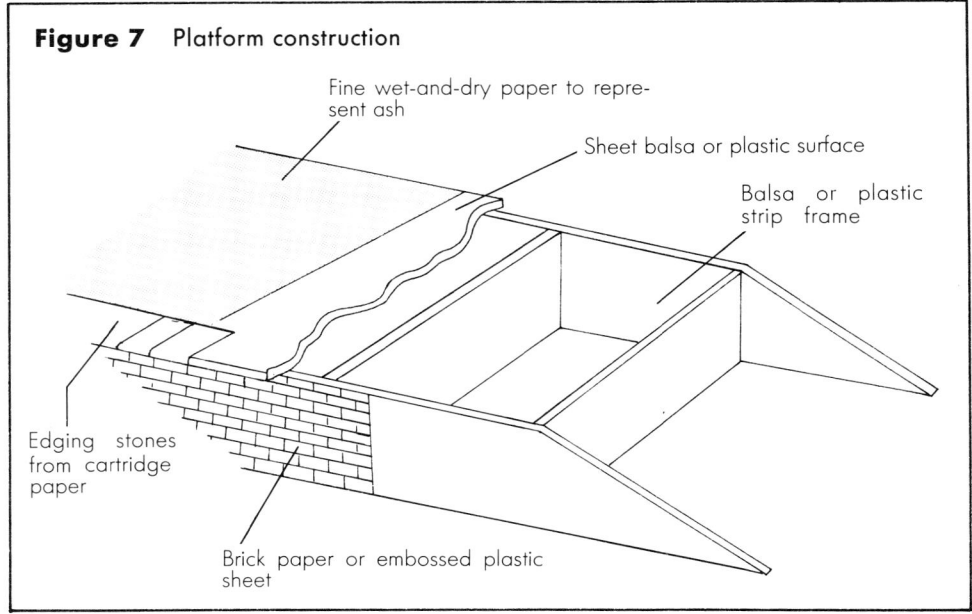

Simple model railway layouts

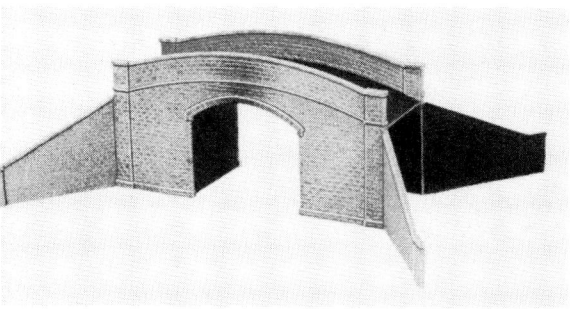

Disguising the fiddle yard often taxes the ingenuity of the layout builder, but Figure 8 shows how a road overbridge and embankment backed with thin plywood or hardboard is used on this layout to provide both an effective screen and attractive scenic feature. This bridge is available from Peco in simple kit form. (Peco Ltd)

its simple construction.

The *Wood-yard* itself is a simple structure, consisting of a small platform, drying sheds, office and perhaps a crane, and can be easily constructed from plastic kits and a Mike's Models yard crane. However, if you prefer not to use kits, it could be constructed quite simply from readily available materials. Building your own has the advantage that it is specifically designed for your layout. Logs can be represented by real twigs and cut timber from sheet balsa or scrap softwood strip. Figure 9 shows the construction method.

Space for any other features on the flat baseboard surface is very limited. However, the area at the front of the

Figure 8 Road overbridge and embankment

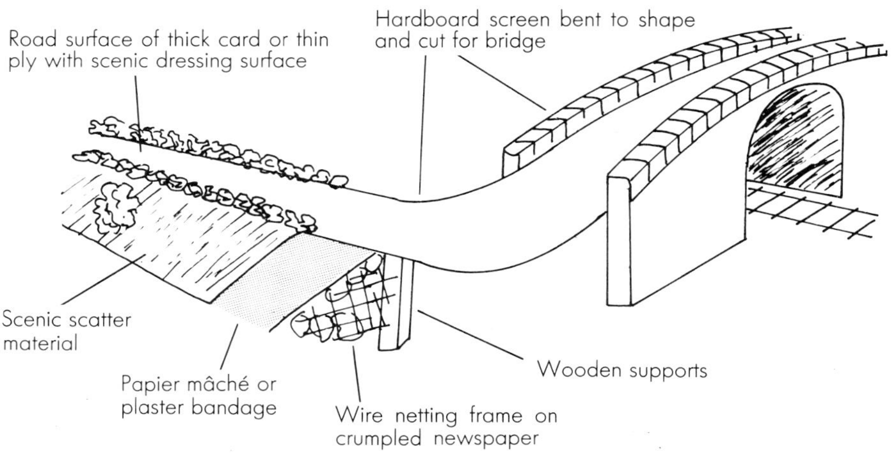

Standard gauge light railway 19

Stage one in the construction of the pond consists of marking on the baseboard surface the area to be 'excavated' and removing the material with a sharp, heavy-duty DIY knife and chisel.

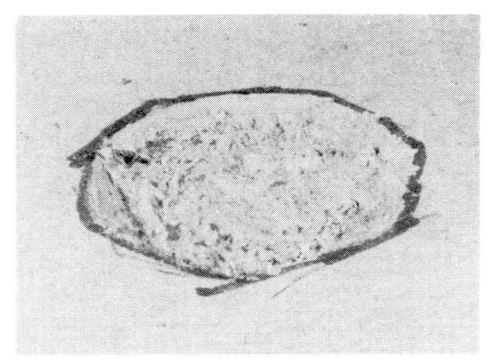

Stage two consists of providing the basic ground cover around the pond and giving the excavation its varnish filling. A shallow excavation can be filled in one go but a deeper one will require several pourings; the thicker the layer of varnish, the longer it takes to dry. It will take a few days in any case, but a skin soon develops. Don't worry at this stage if the varnish seeps into the surrounding area, as this can be dealt with later. Rocks, plants and other details can be added either before the varnish is added or afterwards before it sets. Clear varnish gives a most authentic dirty brown colour, but different colours are available which can create different effects.

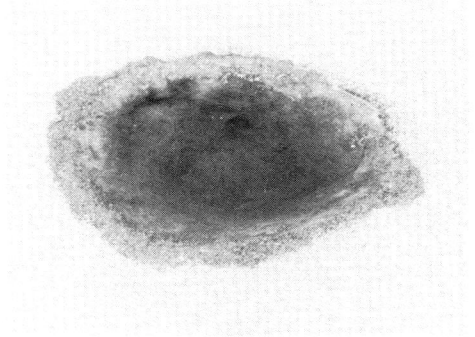

The final stage is adding the detail both to the pond itself and the surrounding area. Here, the texture of the grass around the pond is varied by using different grass mixes, from basic sawdust for finer patches to tufts of foliage mat for coarser weeds. Coarser grasses are also added on either side of the path, beneath the hedge and at the base of the fence. The hedge is rubberized horse-hair covered in various scenic dressings and is readily available. The path and earth banks are created by rubbing off the original scenic dressing with sand paper and painting with a suitable matt colour. The rather artificial-looking tree is from the Britain's range, but with the plastic foliage replaced. Other types of tree and their construction are illustrated elsewhere in the book.

Simple model railway layouts

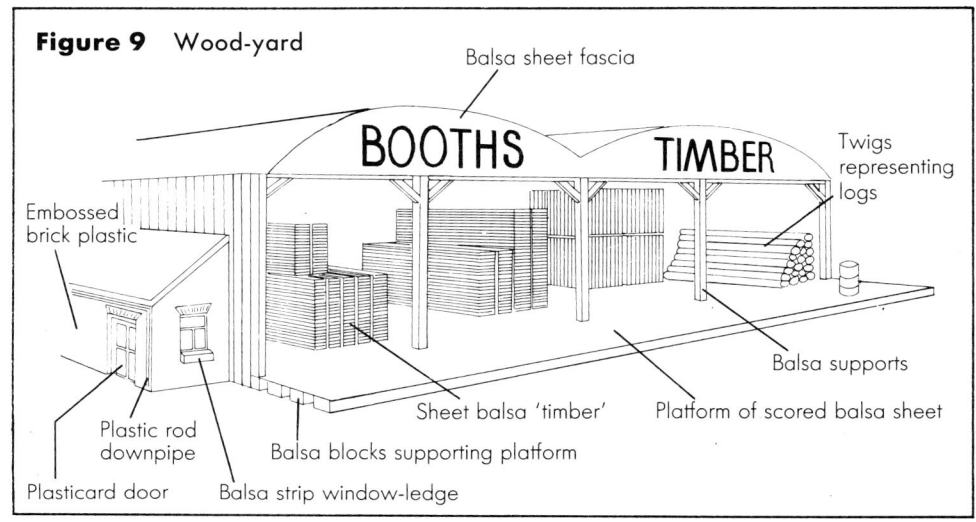

Figure 9 Wood-yard

baseboard could be sculpted out to represent a pond. Paint the depression with a mixture of browns and greys, add some small rocks, a rust-painted bicycle and a steel barrel placed on edge, then fill the 'pond' with a clear varnish to just below the surface of the surrounding ground; before the varnish dries, reeds made from fine bristle and detailed foliage and ferns such as those from the Scale Link range may be inserted. The bank could be painted a brownish-grey earth colour and treated with various shades and textures of scatter material; a path part of the way round could be made by rubbing off the scatter material. A fisherman from the Preiser range or, more likely, two small boys throwing stones, would make a nice 'set piece' (draw the ripples from the thrown stones in the almost dried varnish with the end of a cocktail stick).

The rest of the scenic cover is achieved by simply adding scenic dressings of various colours and textures, remembering that the areas around the engine shed, goods

Although this picture shows a coaling stage on a preserved steam railway, it is nevertheless fairly typical of the small, often improvised coaling facilities provided on light railways. It would be easy to reproduce for this layout using balsa wood or plasticard. The overgrown condition of the track is also an interesting point to note.

Standard gauge light railway

Goods sheds and warehouses come in many shapes and sizes. This is a small pre-cast concrete type, quite common in real life but seldom modelled, and would be ideal for a light railway since it occupies little space. This kit is available from Ratio. (Ratio Plastic Models Ltd)

warehouse and wood-yard should show either ash or a stoned surface, easily represented with the appropriate scenic mix. Don't forget paths and walkways. A group of two or three trees at the rear of the layout between the bridge embankment and the platform would add another eye-catching feature and create some relief; similarly, a single large tree between the embankment and the engine shed would be effective.

Even with a small layout like this, scope for adding and upgrading scenic detail is almost limitless. The best rule of thumb is to look at nature and avoid the impossible — generally, if things don't *look* right, they're not!

Locomotives and rolling-stock

Unless there is a particular prototype in mind when building this layout, the choice is yours, but remember that a twelve-wheel dining-car and an A4 Pacific will totally spoil any attempts to create the atmosphere of a light railway! Such railways were operated

A scene at Tenterden Town on the preserved section of the Kent and East Sussex Railway. A very good 00 gauge model of this ex-Ministry of Supply 0-6-0 saddle tank is available and would be ideal motive power for the layout. Indeed, a 'preserved' line would make an attractive variation on the layout theme suggested.

Simple model railway layouts

A simple engine shed of this type would be ideal for the layout. This one was constructed from a readily available and quite cheap plastic kit and, as can be seen, is quite a pleasing model.

by small locomotives with older, shorter carriages. A few suggestions for commonly available ready-to-run items are set out below.

Locomotives (All tank locomotives)
Hornby: ex-LMS Jinty, LNER J52, Southern M7.
Dapol: GWR 14xx, Austerity J94, ex L&YR 'Pug', ex-LNER J72.
Replica: GWR 57xx

Coaches
Hornby: LNER Gresley brake end, LMS Stanier brake end, ex-GWR Collett brake end, GWR clerestory brake.
Dapol: GWR auto-coach, ex-LMS 57 ft lavatory brake, ex-GWR Collett brake.

The availability of ready-to-run models changes fairly quickly, so check with your local model shop and the modelling press. If any are no longer available, they can probably be obtained on the second-hand market at a reasonable price.

If you can build this, or indeed any other model railway layout, then you can probably build one of the many white metal locomotive kits on the market. Building locomotives from these cast kits is in itself a rewarding hobby, and it does considerably broaden the choice of locomotives available; examples like the Southern P class, the 'Terriers' and the Adams Radials which all worked on light railways become available for the layout. Simple plastic coach and wagon kits are also available and include such useful models as GWR four-wheelers, LNER suburban coaches and a variety of fruit, grain and cattle wagons. Even if you find locomotive building too daunting, these simple kits will add a touch of individuality.

Further References

Britain's Light Railways, Burton and Scott-Morgan, Moorland Publishing

Minor Standard Gauge Railways, R.W. Kidner, The Oakwood Press

Light Railways of Britain, H.C. Casserly, D. Bradford Barton

British Independent Light Railways, Scott-Morgan, David & Charles

The Colonel Stephens Railways, Scott-Morgan, David & Charles

CHAPTER 2
00 gauge branch line terminus

The British branch line railway, and particularly the GWR branch line, has long been a favourite of the railway modeller. The layout plan suggested in this chapter is a straightforward and traditional answer to the space-starved modeller's desire for a realistic model railway in a small space. However, whilst this kind of layout has gained acceptance for model railways, in truth most branch line termini were quite sprawling and often inconveniently situated several miles from the communities they were intended to serve.

There is no such thing as the typical British branch line, as anything more than a superficial glance will prove, and here, probably more than any- where in a large railway company's systems, individuality was maintained. The branches had very distinct features of their own, often having started as small independent concerns.

The plan

Figure 10 shows a simple country terminus station with rudimentary facilities in a space of 8ft × 1ft. The branch line differs from the light railway described in Chapter 1 in that it is altogether a more substantial affair, running in accordance with main line railway practice. There are, however, still restrictions on the locomotives that may be used, basically tank and small tender engines, although in the

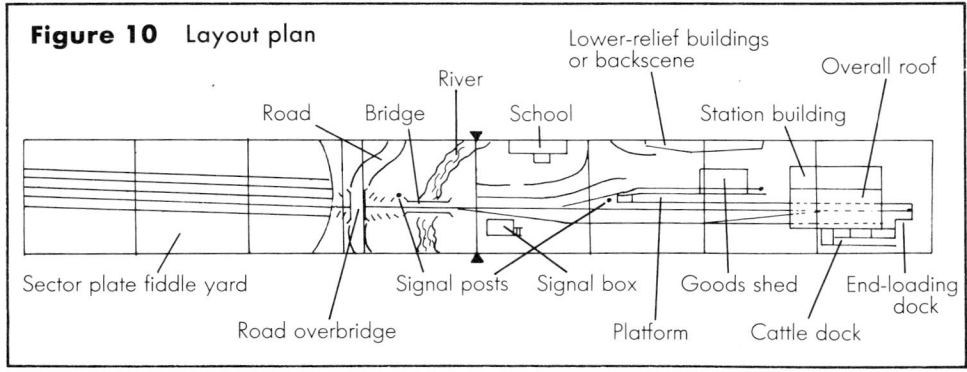

Figure 10 Layout plan

A realistic branch line scene achieved in a narrow space, this 00 gauge layout shows considerable attention to small detail both on the platform itself and in the goods yard area.

last few years of steam on the London Midland Region of British Railways it was possible to see one of the ubiquitous Stanier Class 5 4-6-0's on a small branch line pick-up freight train.

That most popular of all model railway subjects, the Great Western branch line, is depicted here and whilst the following description relates to a GWR line, it can, by changing the obvious company style features, be built to reproduce any other British railway company at any period the builder chooses.

The track plan shown provides for a run-round loop, goods shed and cattle dock; if the layout is modelled as a GWR branch, an overall roof for the station could be provided, a common feature of a number of West Country GWR branch line termini. The run-round loop is quite short and will accommodate just two coaches, or a train of three or four four-wheel coaches if an early (say pre-1930) period is chosen for the model.

Even though the station area occupies an area only about 5ft in length, the variety of goods that can be accommodated, the goods shed, coal staithes, cattle dock and end-loading facility all provide plenty to occupy the operator in shunting the station. In all probability the passenger service would be by autotrain (probably the ubiquitous 14xx and autocoach) or railcar. Extra interest can be provided by adding milk vans and shunting these to the end of the platform for loading and unloading. If a West Country setting is

00 gauge branch line terminus _____ 25

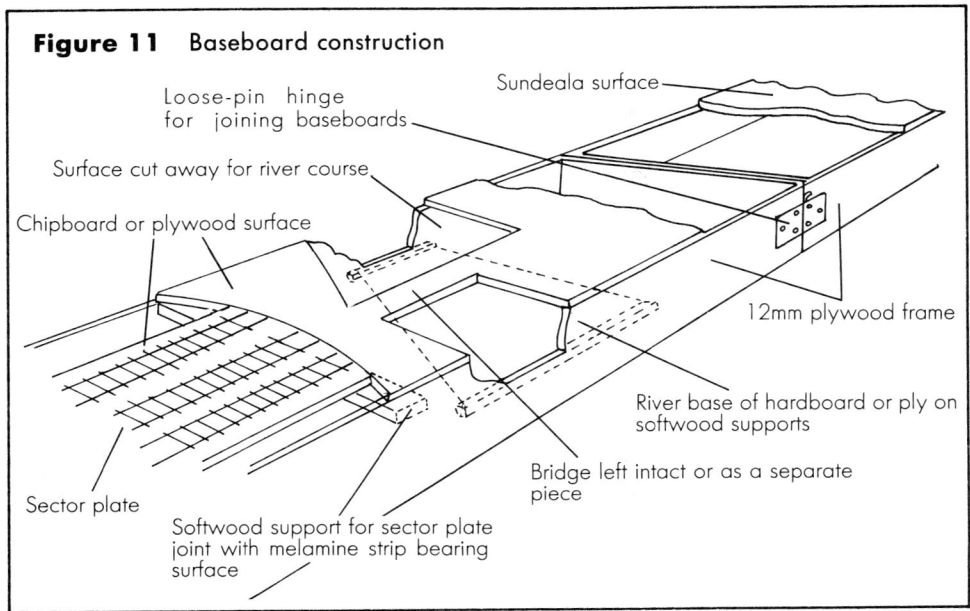

Figure 11 Baseboard construction

chosen, a through coach or an additional coach to meet peak holiday demand could be run from time to time.

Baseboard construction

The general principles of baseboard construction discussed in Chapter 1 apply also to this layout. Two 4ft × 1ft baseboards of plywood and a Sundeala-type material are again needed. However, there are two new techniques required for one of the baseboards involving a dropped area to provide for a stream flowing beneath the railway as it approaches the station and the construction of a 'sector plate' fiddle yard.

Dealing with the simpler of the two baseboards first, that on the right carrying the station itself, this is a straightforward box of 12mm ply framework covered with a surface of Sundeala or similar board. The left-hand baseboard, on the other hand, needs to accommodate the dropped section for the stream and to provide for the sector plate, and Figure 11 shows the arrangements for both features as well as general construction methods.

The sector plate arrangement is useful in that it enables a multi-road fiddle yard, representing the rest of the railway system to and from which the trains operate, to be accommodated in a reduced space by removing the need for points into each of the sidings. A normal set of three hidden sidings would require either two ordinary points or a three-way point, and thus a minimum of 12in is saved by using neither. It is a simple arrangement, consisting of a track bed carrying the required number of roads or sidings pivoted at one end and curved at the other; the adjacent fixed board is curv-

The very authentic appearance of SMP 00 gauge trackwork is well illustrated here, and there are also finely-detailed, easily-assembled, plastic-sleepered point kits available up to a 36in radius. A considerable number of more complex pointwork formations are available using the well-proven method of soldering rail to copper-clad sleepers, and a number of advertisements appear regularly in the model press offering these points ready assembled at very reasonable prices.

ed to match. This enables the fiddle sidings to be swivelled, allowing each of the roads to be lined up with the main running line. This curved 'joint' must be even and clean and is easiest to obtain if both the pivoting 'table' and the fixed portion of the baseboard are cut from the same piece. Mark out the arc on a radius from the pivot point. Iron-on melamine strip, available from DIY stores, is fixed to the cross members beneath the sector plate and acts as a bearing surface. The surface of this baseboard should be ½in plywood, as Sundeala-type board is not sufficiently rigid or durable.

The baseboards are joined together by brass hinges; the pin head is filed off and the pin removed to be replaced with a brass rod of similar diameter to make an easily-removable locating pin.

All joints should, of course, be glued, pinned or screwed for strength.

Trackwork

Trackwork representing a British branch line should be of bullhead rail. There are now a number of readily available bullhead rail flexitrack systems on the market, not only for 00 gauge 16.5mm track but also for the more accurate EM 18.2mm gauge and S4 18.83mm gauge (the correct track gauge for 4mm scale models). Both the SMP Scaleway and the Alan Gibson track systems provide compatible point kits which are easily assembled, and provide points with an appearance far superior to the commercially-available flat-bottom rail varieties. Either of these track systems would be ideal for this layout and the point kits contain easy-to-follow detailed instructions. Figure 12 shows the main components of and terminology commonly used for turnouts or points.

Before the track is laid it is advisable to position it on the baseboard to ascertain the best position for the pointwork, giving maximum clearances and the longest possible sidings and loop, and, particularly important, to avoid placing points across the baseboard joints or locating point motors where

00 gauge branch line terminus   27

sub-frame structures would interfere. Draw the centre line of the final trackwork position with a marker pen.

Lay the cork sheet underlay where the track is to go on the baseboard surface. Cork sheet is available in rolls from most model shops, although strips ready cut and bevelled in half-track widths ready to be joined on the track centre line are now available from specialist retailers advertising in the model press, and are easier to use.

Pin the track lightly to the baseboard, starting with the pointwork and cutting and curving the plain track to fit in between. It is much easier to adjust the alignment of plain track to fit pointwork than vice versa. Avoid any kinks or sudden changes of direction, and try to achieve smooth curves, particularly at rail joints.

Now carefully lift the track, spread the area where it is to be laid with an even coat of PVA woodwork glue. The areas where the point tie bars are to slide should be marked and left free of glue. The track can now be relaid and pinned; check the alignment very carefully again and make any final adjustments. Carefully spread the ballast over the track and leave to dry. If you're not confident about achieving a straight or even edge to the glue where the ballast is to end, use masking tape.

The trackwork should be thoroughly tested before proceeding further. (See the Scenery section below for the method of carrying the track on the bridge over the stream.)

Wiring

The basic wiring requirements for this layout are extremely simple. Figure 13 shows the basic wiring using two switched feeds. A switched dead end sec-

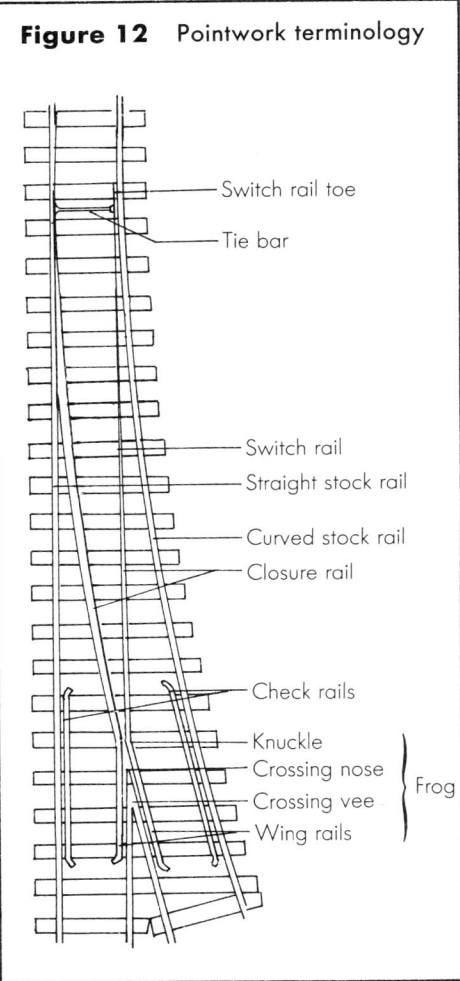

Figure 12   Pointwork terminology

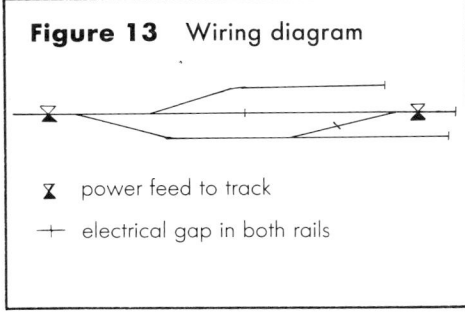

Figure 13   Wiring diagram

    X   power feed to track

    +   electrical gap in both rails

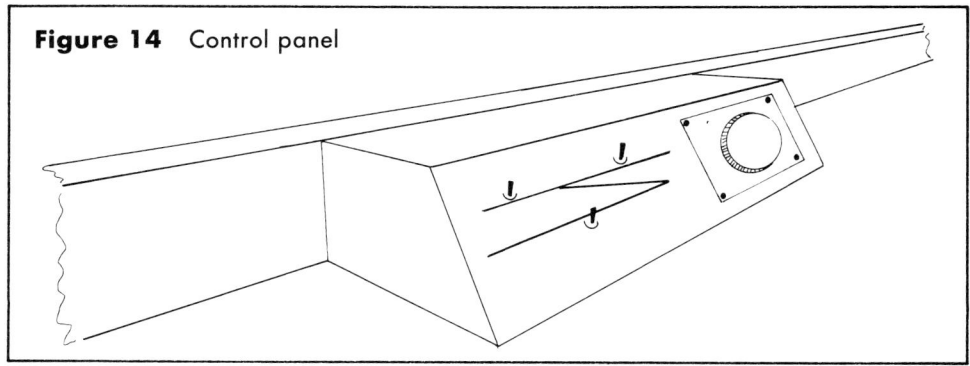

Figure 14 Control panel

A multi-pin plug and socket. These are readily available from model shops and electronic parts suppliers. A number of specialist suppliers of electrical parts that are suitable for model railway use regularly advertise their products, ranging from simple switches through to complete control systems.

tion could be added in the spur of the run-round loop to allow for locomotive storage. If point motors are used, the wiring arrangements recommended by the manufacturer should be followed.

The wiring for both the track feeds and the point motors should be harnessed neatly under the baseboard and run to a multi-pin socket. The control panel, shown in Figure 14, should be removable so that the layout can be operated from either the front or rear; the wires from the panel should be harnessed together and run to a multi-pin plug which connects with the socket in the baseboard.

The wiring arrangement for the sector plate fiddle yard is simple and is shown in Figure 15. The adapted door bolts used here serve not only to align the track but also to switch power to the selected fiddle yard road.

Scenery

This small layout can accommodate quite a large amount of scenery and offers good scope for both model buildings and the landscaping associated with the stream and bridge.

The *bridge* carrying the railway over the river can be made from a simple kit in the Wills scenic range (see the photograph). The decking which carries the track is made from either thick plasticard or thin plywood which slots between the stream banks. Embossed stone retaining walls for a short distance on either side of and beneath the bridge give the structure engineering credibility (these are included in the kit).

The *river* itself is made very simply as illustrated in Figure 16. The

00 gauge branch line terminus ───────────────────── 29

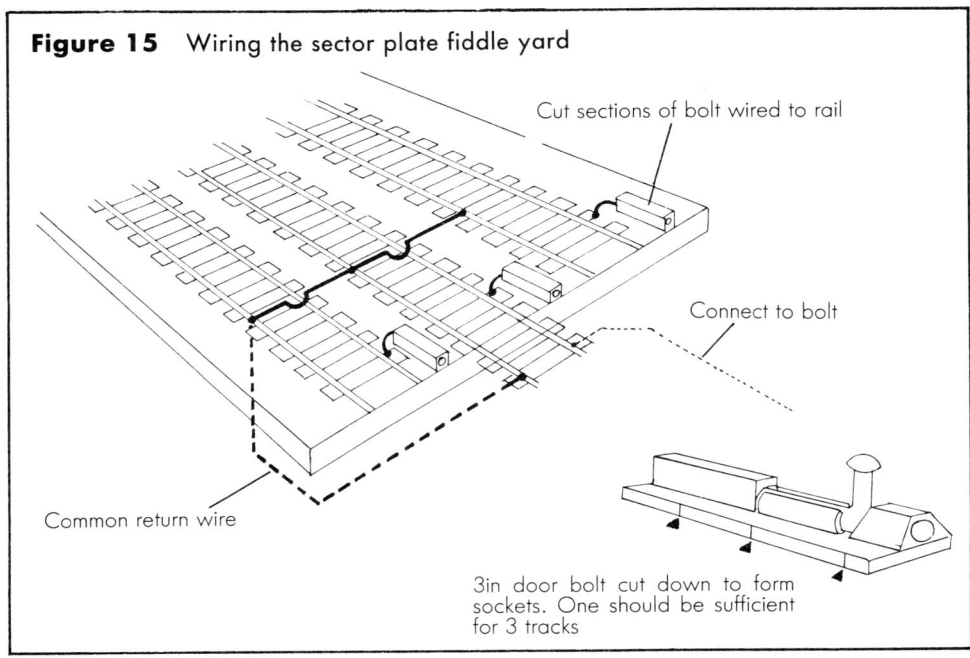

Figure 15 Wiring the sector plate fiddle yard

Cut sections of bolt wired to rail

Connect to bolt

Common return wire

3in door bolt cut down to form sockets. One should be sufficient for 3 tracks

baseboard surface is cut along the curved course of the river and a section about 2in wide is removed; avoid too regular or straight a cut because nature doesn't follow straight lines! Narrow the cut towards the rear of the baseboard to help give an illusion of distance and make it easier to blend it into the backscene. The actual surface or bed of the river is a piece of hardboard or similar, an inch or two wider than the gap, fastened beneath the baseboard surface and supported at either end by a piece of 1in × 1in softwood screwed inside the baseboard mainframe. The resulting 'banks' in the mainframe should be shaped and carved roughly with a knife or chisel to

The Wills bridge kit referred to in the text. The small river is very similar to the type suggested for this layout and shows how effective ordinary household varnish can be in representing water. (Author's collection)

Simple model railway layouts

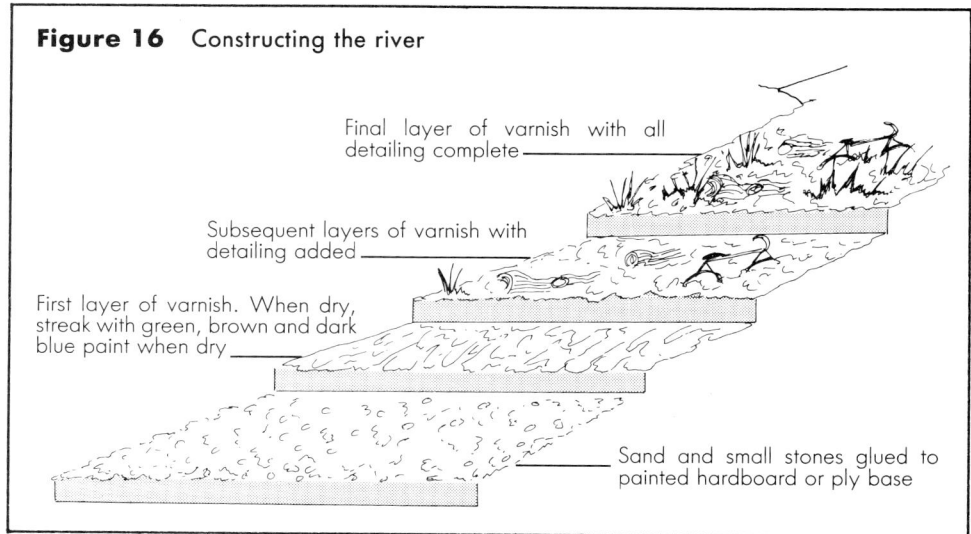

Figure 16 Constructing the river

- Final layer of varnish with all detailing complete
- Subsequent layers of varnish with detailing added
- First layer of varnish. When dry, streak with green, brown and dark blue paint when dry
- Sand and small stones glued to painted hardboard or ply base

create the desired effect, remembering that the outside bank on a curve will often be steeper because the water is cutting into the bank, while the inner bank will be shallower.

The stream bed can be coloured and textured in a variety of ways. Paint it in streaks of browns and greens and perhaps glue down a bit of sand. Small rocks can be carefully added and the area immediately 'downstream' of them painted white to represent the foam of broken water. If in doubt, have a look at a real stream. The water itself

Another view of the Wills bridge shows how the river can be curved into the backscene to disguise an obvious end at the edge of the baseboard — a small road overbridge helps with the disguise. Further details have been added round the bridge, including areas of dense undergrowth made from foliage mat; the platelayers' hut, complete with grind wheel and platelayers tools, is from Cooper Craft. (Author's collection)

00 gauge branch line terminus 31

can be represented by carefully pouring clear gloss varnish over the bed, not too much at a time; several good even coats are required. Reeds and other vegetation can be made from bristles, wire wool and scenic foliage, or, for a super detail model, individual ferns and rushes from the Scale Link etched range. A rusty bicycle frame (a commercial plastic bicycle moulding with the wheels cut off and the frame painted a rust colour) embedded in the foliage adds character. After several days drying, add further coats of varnish in a similar way until the desired result is obtained.

The stream banks can be treated with sand, rocks carved from Polyfilla and scenic foliage and grass materials to obtain the desired finish. A fisherman or animal drinking from the water adds character and detail, or what about some swans or ducks from the Dart Castings range of scenic castings?

Areas to be 'grassed' as rough pasture can be created by firstly covering the desired area with a coating of dark brown or black matt emulsion then coating with a mid-green grass material when dry; build up the desired textures with coarser turf and foliage mixes. Create a path by rubbing away the grass covering and coating with a matt earth brown mix; fine sand or ash

Right *A path is rubbed into the ground cover with sandpaper and coarse grass is added on either side. Be sure to paint the path using matt colours.*

Below *For those who don't want to make their own platforms, a simple but effective modular system is available from Peco. The individual units can be joined to provide any length or width of platform desired* (Peco Ltd)

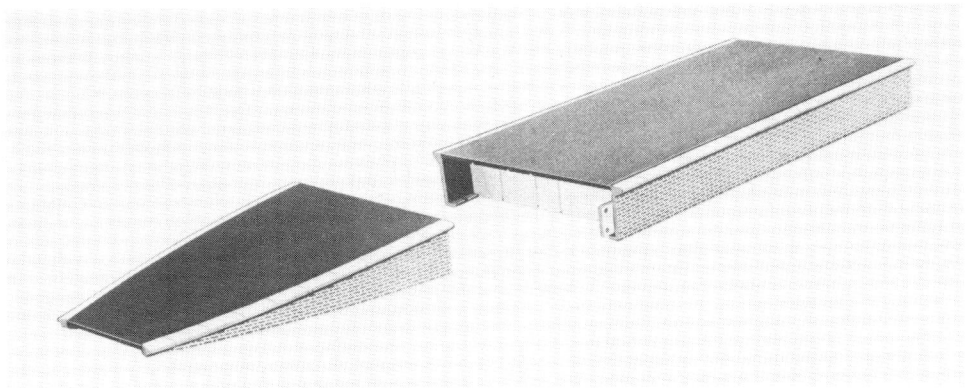

scattered here and there can be effective. Remember to add coarse grass near the edge of the path or around tree stumps (easily made from twigs found out walking or in the garden), fencing posts and walls.

The *platform* is easily constructed from strips of wood and the edge can be covered with a proprietary embossed ready-coloured stone card. Faller produce several varieties and colours; choose the colour which suits the stone of the area in which your model is supposed to be set. A platform kit system, readily adapted to suit individual locations, is available from the Peco range and is another alternative.

The *station buildings* of the Great Western branch line have featured prominently in the model railway press over the years and there are a great many scale drawings to be found in the various publications. Your local library should have or be able to get the Leleux Index of Drawings which is a good place to look for any that have appeared in magazines. If you haven't got the magazine or access to back numbers, the library can get you copies from the British Library for your own personal use at a very small charge. Ian Allan have recently published a similar drawings index. From the plans you obtain you will be able to scratch build suitable models for your layout from plasticard or even cardboard to give it that air of individuality. Scratch building your own structures is relatively simple and the bibliography at the end of this chapter gives some references to good books on this aspect of the hobby which in itself can become quite absorbing.

Alternatively, there are kits, both plastic and cardboard, of specific prototypes, usually Great Western. Hornby, for example, recently produced Dunster Station which, with the replacement of the self-adhesive 'stone' paper by embossed plasticard, can make an effective model. The Peco range also includes station building kits which are simple to construct but are not modelled on specific prototypes.

An excellent kit for a GWR station building, ideally suited to this layout, is available from Ratio. (Ratio Plastic Models Ltd)

00 gauge branch line terminus _____ 33

Figure 17 Station overall roof

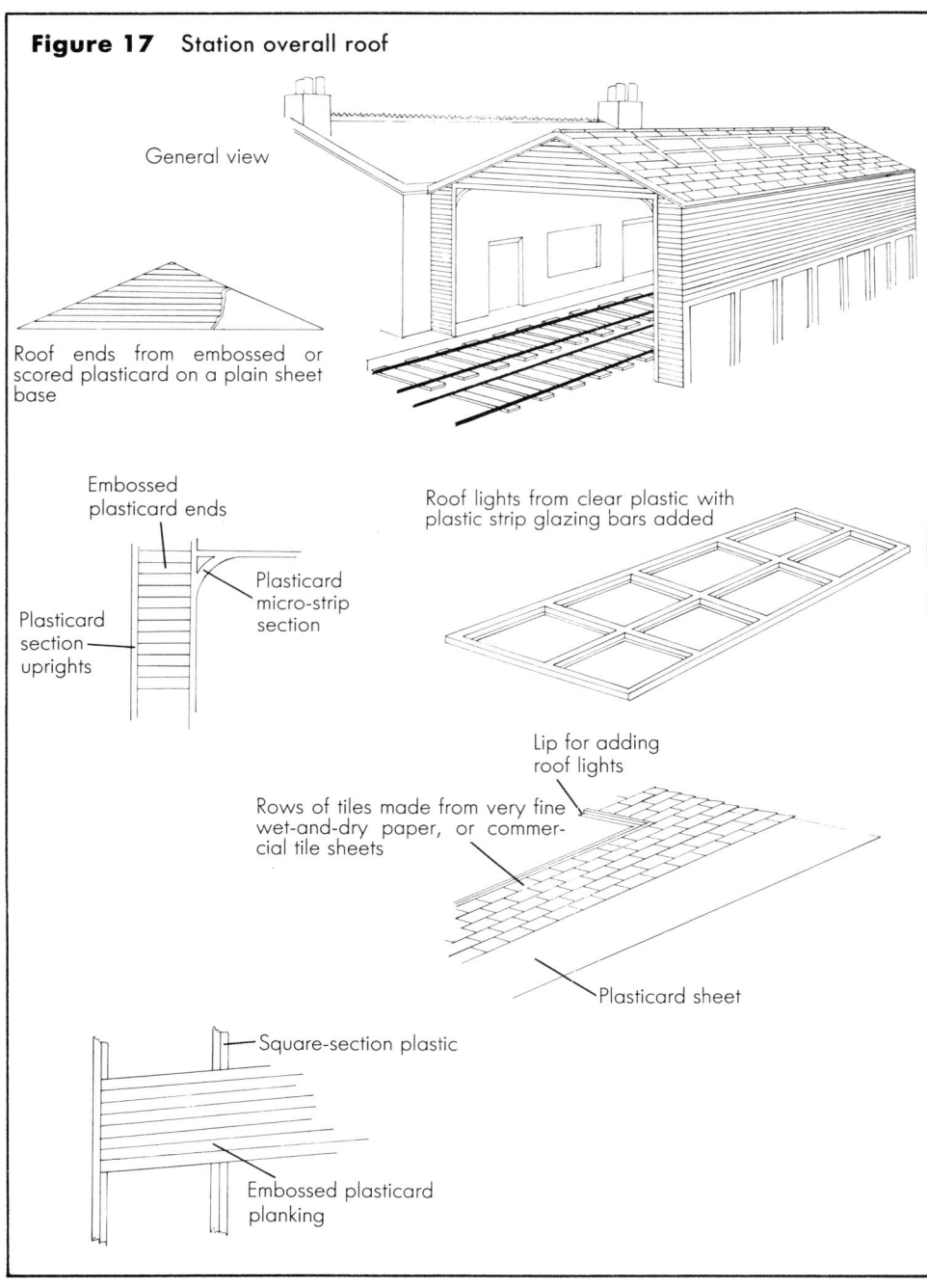

34 — Simple model railway layouts

The cattle dock was once a common feature of many British railway stations, and was used for the loading and unloading of livestock, once a very common commodity on our railways. The cattle dock shown is made from the highly-detailed Ratio plastic kit and is shown in a typical rural setting. Points worthy of note are the grounded van body and its surrounding detail, and the importance of bedding in buildings and structures to avoid unsightly gaps at the base of walls; the dock and van body look like long-established fixtures rather than temporary additions. This is all part of creating a realistic scene. (Ratio Plastic Models Ltd)

Ratio produce an excellent and detailed station building kit in plastic modelled on an actual GWR prototype. If desired, the overall station roof can be constructed relatively simply from plasticard as in Figure 17, and adds a distinctive feature.

The *cattle dock* is built up in the same manner as the station platform, and fencing, suitably adapted, can be used for the cattle pens (don't forget to add some cattle!). Alternatively, use one of the easy-to-assemble plastic kits.

Ratio produce an excellent small GWR *signal box* which is very detailed and ideal for this layout.

The space between the stream and goods shed area at the rear of the layout could be occupied, perhaps, by a school and playground as shown in Figures 18 and 19. This would form an interesting feature in itself and could be easily constructed using the Dapol church as a base. Use some of the many

A typical small Great Western signal box, showing much excellent detail. Given a realistic setting, the box itself can be further enhanced by the addition of full interior detail for which Springside produce a kit. The picture also shows other important details which help to create a realistic layout, including a GWR signal, platform fencing, post and wire boundary fencing in the background and different types of ground cover. The trackwork is a fine scale bullhead rail, very effective in portraying the right kind of track for a branch line layout. Point rodding and signal wire from the box could be added — parts are available from a number of sources. (Ratio Plastic Models Ltd)

00 gauge branch line terminus _____ 35

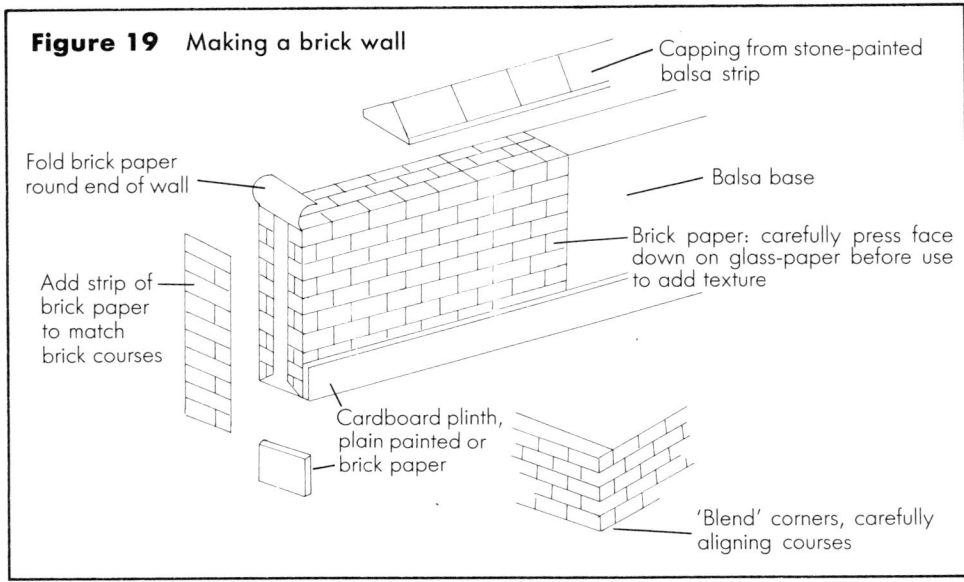

Figure 18 The school

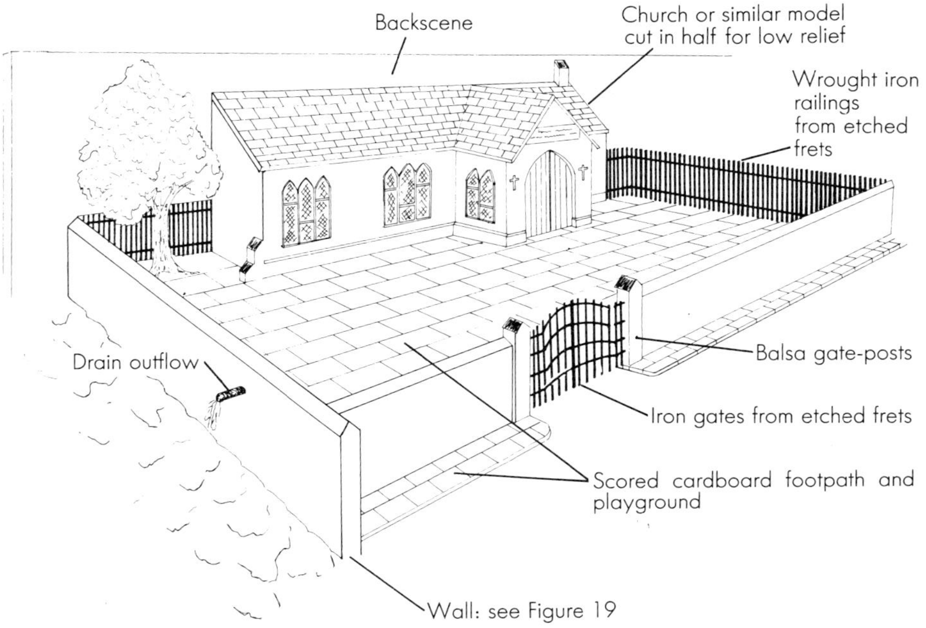

- Backscene
- Church or similar model cut in half for low relief
- Wrought iron railings from etched frets
- Drain outflow
- Balsa gate-posts
- Iron gates from etched frets
- Scored cardboard footpath and playground
- Wall: see Figure 19

Figure 19 Making a brick wall

- Capping from stone-painted balsa strip
- Fold brick paper round end of wall
- Balsa base
- Add strip of brick paper to match brick courses
- Brick paper: carefully press face down on glass-paper before use to add texture
- Cardboard plinth, plain painted or brick paper
- 'Blend' corners, carefully aligning courses

excellent figures available to populate the playground, grouping the figures realistically, perhaps playing football or hopscotch. Mark 'goals' on the inside of the walls and hopscotch grids on the ground. The school building itself should be modelled in low or half relief to give a greater illusion of depth; in other words, only the front half of the building is modelled, backing on to the backscene.

The backscene covers the rear of the model and not only provides a frame for the layout but also, painted a pale blue/grey or having sky paper pasted on to it, can help to create an illusion of distance, particularly if models are blended into it.

Even though this is only a small layout, a great deal of detail can be added, limited only by your imagination and powers of observation, which can add that final intangible dimension to a model. Use the excellent and relatively cheap detailing packs available; mention has already been made of children in the playground and cattle in the cattle dock, but what about dustbins, street lamps, telephone kiosks, crates and packages, people and even dogs and cats? Period road vehicles are available from a variety of sources, from cheap plastic models to well-detailed and more expensive kits — everything from a Model T Ford to a modern articulated lorry.

Ratio produce excellent working GWR signals (and also SR and LMS/BR types) which help to bring the layout to life. Cast runners are available for point rodding which can be made from the thin wire sold in straight lengths for locomotive handrails. Representation of cranks can also easily be made.

The scope for detailing is endless and is a very enjoyable pastime in itself. Many a happy hour can be spent arranging groups of figures or animals until you have them 'just right'.

Locomotives and rolling-stock

The popularity of the GWR with railway modellers has ensured that it is the best catered for of all the British railway companies for models of locomotives and rolling-stock. If all the various kit and ready-to-run ranges are considered, there cannot be many GWR locomotive classes or items of general rolling-stock not produced for the 4mm scale modeller.

The ubiquitous pannier tank is available both in ready-to-run 00 gauge and from kits in most of its variations. The 14xx autotank is similarly available as are two eminently suitable small tender locomotive prototypes, the 0–6–0 Dean Goods and Collett Goods.

Lima produce an excellent diesel railcar; this and the steam locomotives can be easily detailed further by using the many detailing and flush glazing window packs available.

An autocoach, or, to be more precise, one of its variants, is the subject of an excellent ready-to-run model available in both GWR and British Railways liveries. Indeed, most of the ready-to-run models are available in liveries of various periods, usually from the 1930s to the 1950s.

There is plenty of choice in ready-to-run goods rolling-stock including even specialist vehicles such as the Siphons G and H. If the various kits are considered, ranging from simple plastic models to the more complex etched brass, then vehicles suitable for all

00 gauge branch line terminus 37

There are a great many kits of varying complexities available for GWR locomotives. K's 44xx is a key type for a West Country branch line and, as can be seen from the photograph, it makes up into a very respectable model. (N. & K.C. Keyser Ltd)

periods from the early days of the GWR (even broad gauge,) through to the end of the steam era are available.

If you are prepared to consider building kits for your locomotives and rolling-stock, itself a fascinating and rewarding aspect of the hobby, then almost any period could be chosen in which to set your layout. The 517 class 0-4-2, predecessor of the 14xx,

The Lima GWR diesel railcar makes a contrasting alternative to the more common 14xx autotrain for a layout set in the late 1930s or later. This model is available in both GWR and BR liveries and can be detailed further with the addition of flush glazed windows, etched windscreen wipes, vacuum pipes and turned brass buffers. Wire handrails near the doors can be fabricated from handrail wire or small wire office staples.

As mentioned in the text, the Great Western Railway did not have a monopoly on branch lines and you can easily choose to build a layout representing your favourite company or period. The M7 was common motive power on Southern Railway branch lines and Hornby have produced this model of the class (Hornby Hobbies Ltd)

becomes a possibility — an excellent kit is available from M&L Models' Premier range. Similarly, kits for older coaches, such as Ratio's plastic four-wheelers, and older goods vehicles, from D & S Models and other kit manufacturers in cast and etched metal, broaden the scope of the period which could be modelled.

Further References

Miniature Building Construction, J.H. Ahern, MAP

Architectural Modelling, D. Rowe, Wild Swan

GW Branch Line Termini (combined volume), P Kareau, OPC

Modelling the GWR, C. Ellis, Ian Allan

CHAPTER 3
00 gauge 'modern scene' layout

The layout described in this chapter is one which is often referred to in the model railway press as a 'modern image' layout. So often the current railway scene is considered inappropriate for small layouts, assumed to be only for the carriage of people and goods en bloc in long commuter and InterCity trains and block freight trains, offering little scope for the space-starved modeller. Happily, there are exceptions to this general view, although they are tending to disappear rather quickly. It is hoped that this layout project will show that a realistic, believable model with good operating scope can be built to represent the contemporary scene in the limited space of the modern house.

The plan

It is intended that this layout should depict the remains of a railway terminus in a large northern town. The terminus would, when built in the Victorian era, have reflected the prosperity of the town and the faith in the future of both the area and the railway company. However, in the last quarter of the twentieth century after changes in the method of railway traction, the economic importance of the railway, the community it serves and the growth in alternative transport not envisaged when the railway was built, the station has suffered a good deal of modernisation, rationalisation of operations and facilities and has had a good deal of 'amputation' along the way. It is assumed that the town, whilst no longer requiring the services of a major passenger station, still has sufficient economic life to justify the retention of some kind of railway presence.

The station's passenger facilities have been reduced and part of the former station area is now occupied by a British Fuel coal concentration and distribution depot. The large, former railway warehouse close to the station remains, but today provides the headquarters and distribution centre for a large mail order company.

Whilst the station itself no longer boasts InterCity passenger trains, it has a passenger service which links the town to InterCity trains at a station some 10 or 12 miles away. There is also a local service to a number of surrounding towns which still have rail links.

This scenario gives a fair degree of variety in operation and, of course, re-

Modern commercial or industrial development encroaching on former railway land and abutting on to the remaining running lines is a common feature. The building to the right of the picture is the rear of a large superstore which now occupies the site of a former goods warehouse not unlike the one suggested for the mail order depot. The picture also shows a station building which is carried on a bridge over the railway. The old loading gauge still remains but with the sidings removed, providing a suggestion for added detail.

quires rolling-stock to undertake it, a good excuse to take full advantage of the excellent models available.

Passenger services to the station would be in the hands of ageing diesel multiple units. Here some variety could be introduced as there are two distinct services, the local one and the one to the InterCity station. In reality, these services might be operated by the same class and formation of DMU, but over the last few years in the area in which the layout is set, as the ageing fleet of DMUs has declined in operational availability, all sorts of combinations of type, sets and, not least, livery of trains have appeared. Two-car sets would be the usual formation, perhaps strengthened at peak times to cope with commuter traffic.

The freight services would again consist of two distinct operations, one serving the mail order warehouse, the other the coal depot. Whilst this latter operation may seem dull and uninteresting compared with the variety associated with the steam age 'pick-up' goods, the nature of the facilities does provide a degree of visual openness which complements and contrasts with the general surroundings of the layout and the mail order warehouse. (it is important to avoid claustrophobia in a small layout). It also helps to illustrate and emphasise how the old railway has been adapted to meet current needs with usually little regard for aesthetics! The basic operation associated with this facility would be the arrival of coal by rail, which would then be loaded by mechanical shovel and conveyor belt into hoppers for local distribution by road. The effect could be equally well attained with bulk cement, roadstone or a small-scale container transfer facility.

The mail order warehouse would require 'feeding' with goods, and parcels would be sent out, it is assumed principally by rail, in response to orders.

00 gauge 'modern scene' layout

Figure 20 Layout plan

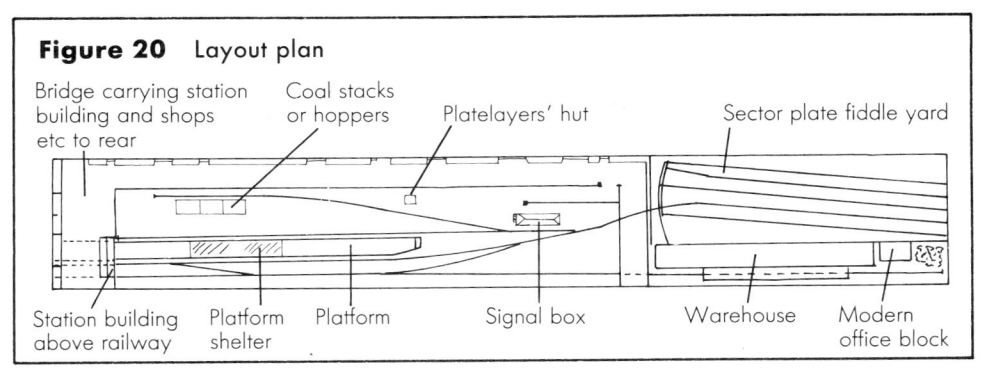

- Bridge carrying station building and shops etc to rear
- Coal stacks or hoppers
- Platelayers' hut
- Sector plate fiddle yard
- Station building above railway
- Platform shelter
- Platform
- Signal box
- Warehouse
- Modern office block

The type of vehicles used would be varied, but would all be fitted covered stock. This would include various passenger brake vans of pre-nationalisation origin (in particular the LMS 50ft steel-sided type, which continues to make regular appearances on such services, and the ex-Great Western Siphon G). The British Railways equivalent, the short wheelbase fitted goods vans, still a very common sight, and its modern, all steel, long wheelbase version, not forgetting the ubiquitous CCT, would also be used. Plenty of variety here, particularly if the varied liveries and state of cleanliness of these vehicles are taken note of. Whilst the simplest form of operation would be one train in and out per day, using modellers' licence (imagination is a wonderful thing!) it would be possible to bring in the odd additional vehicle from time to time.

Now that a believable scenario for the layout's existence and an idea of the operations have been evolved, it is time to consider the design of the layout itself. There are a number of points which will determine the final layout plan, and these can be summarized as follows:

1. The layout is a terminus and therefore the trains need to be seen to run to and from the rest of the British Rail network. For the purposes of the model, this is represented by a fiddle yard, a small number of sidings divorced from the main part of the layout where trains can be made up and marshalled. Because it is separate and screened from the operational and scenically developed part of the layout, trains will appear to go somewhere and arrive from somewhere in a realistic fashion.
2. Passenger facilities are required for our DMU services. Two platforms would be ideal, but one, with careful timetabling of trains, would suffice.
3. There is a coal distribution depot.
4. There is a former railway goods warehouse, now a mail order depot.
5. As already stated, the layout represents a 'rationalized' station with a proud if rather grandiose architectural heritage. It is situated in a large provincial town in the north of England and care must be taken not to forget this in developing the scenic side of the layout, particularly the buildings, back-

ground and civil engineering. However, perhaps the most significant design restriction is the space which is likely to be available for the completed layout. This will also depend on whether the layout is to be permanently erected or be portable and normally stored when not in use. This layout is designed to be suitable for either situation, so if necessary it could easily be transported in an average family car and used for public exhibition.

Bearing in mind the design criteria we have evolved, and the type of operation envisaged, a layout 10ft 6in × 1ft 6in is suggested; this will meet those criteria and allow the use of trains consisting of one diesel locomotive and up to six vehicles and DMUs of up to three cars in length. The suggested layout will be seen in Figure 20.

Baseboard construction

The baseboards needed for this layout are conventional in design and follow closely the principles detailed in Chapters 1 and 2. The road overbridge and the 'station below road level' effect need to be incorporated in the design and construction of the baseboard, and the backscene, rather than merely representing sky or open countryside, should give the impression of a major town. For added depth, the retaining wall surrounding the station area should have some representation of a roadway between it and the backscene (Figure 21).

Three baseboards are suggested, each 3ft 6in × 1ft 6in. If sufficient space is not available, the length and width could be slightly reduced, but train length and scenic features would be reduced accordingly. An alternative would be for the baseboard along its length to end at the retaining wall made 6in or so high above baseboard level and for the road overbridges to remain at the extreme left-hand end, incorporating a station building and at the fiddle yard end. The baseboard should be made from 12mm plywood, with Sundeala or a similar type of fibreboard

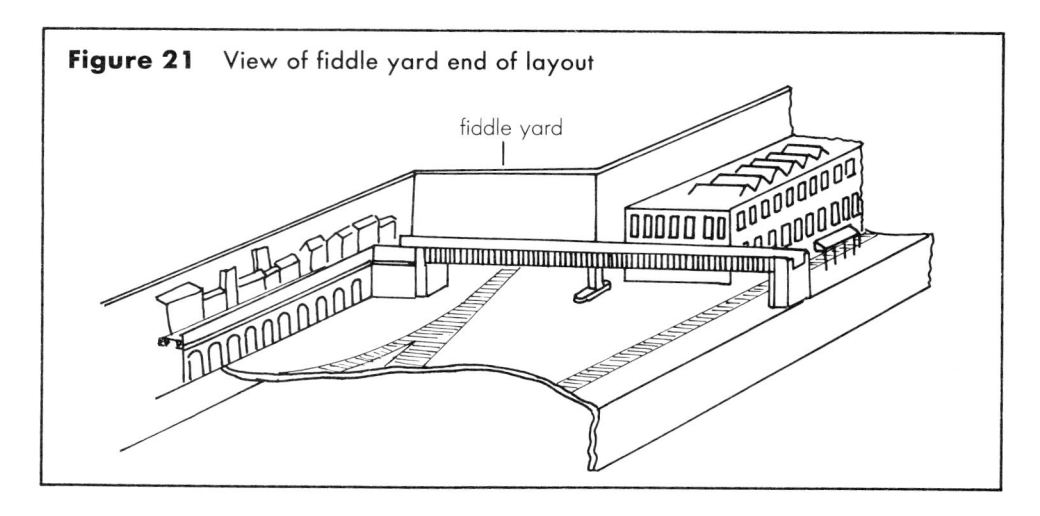

Figure 21 View of fiddle yard end of layout

00 gauge 'modern scene' layout _____ 43

Figure 22 Sector plate fiddle yard

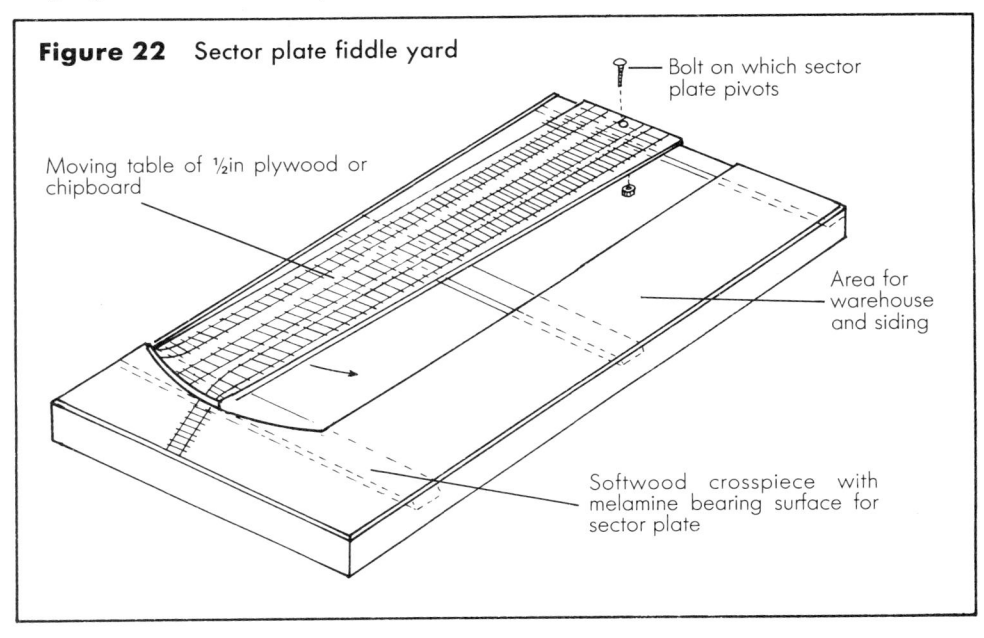

- Bolt on which sector plate pivots
- Moving table of ½in plywood or chipboard
- Area for warehouse and siding
- Softwood crosspiece with melamine bearing surface for sector plate

for the surface. The fiddle yard arrangement shown on the plan is of the space-saving 'sector plate' design, different from the traditional 'ladder' arrangement by not requiring a point or turnout for each fiddle yard road. However, it needs to be incorporated at the baseboard construction stage. The top surface of the fiddle yard baseboard is made from plywood of the same thickness as the other baseboard surface material. One end of the portion of board incorporatating the fiddle yard is cut in an arc radiating from the pivot point and pivoted to the framework at the other to allow the yard to be swivelled and individual roads lined up to allow the entry and exit of trains to and from the station. It sounds complicated but is really quite a simple carpentry exercise. Figure 22 illustrates the construction, and more detail is given in Chapter 2.

Of course, a traditional fiddle yard using, say, two points to give the three roads could be used but only by sacrificing the length of the sidings and therefore train length.

Trackwork

Trackwork for a contemporary station of this type would probably be mainly flat-bottom rail on wooden sleepers, with possibly some older bullhead rail remaining in sidings and some newer concrete-sleepered track in places.

Whilst there is some good bullhead rail trackwork available with easy-to-build matching point or turnout kits, there is currently no model equivalent of the more modern flat-bottom rail required. However, a kit system from Peco is now on the market and it is well worth considering this if truly realistic 'fine' track is desired. Track building is not really difficult and can be well

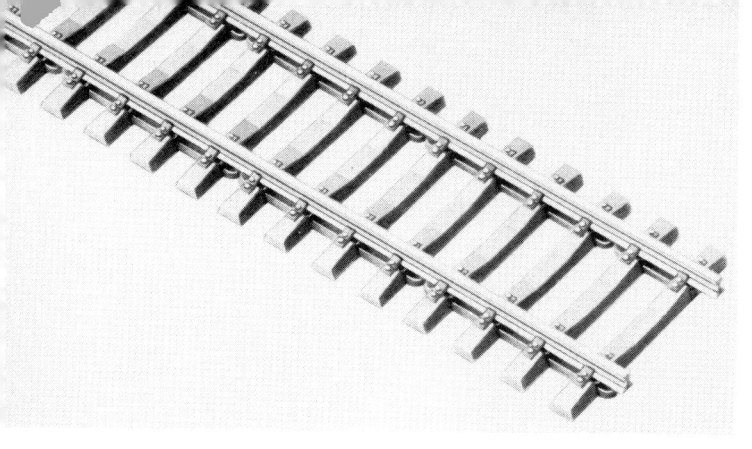

Concrete sleeper trackwork is available from Peco in both 00 gauge and N gauge and would be suitable for a modern image layout. (Peco Ltd)

worth the effort. However, if track building is not for you then the only alternative is to resort to the Peco Streamline range of flexitrack and points. It does not represent British flat-bottom rail trackwork to exactly 4mm scale but it does come complete with ready-to-lay points and crossings, point motors, switches and a whole range of accessories. It is perfectly acceptable, readily available and of good quality, and it is available with both wooden and concrete sleepers. Points are available with both live and dead 'frogs'; the dead frog points are easier to wire but better running is achieved with live-frog pointwork.

Tracklaying using this system is straightforward but it is still a good idea to play about with lengths of track and pointwork on the baseboard first, to establish the optimum position bearing in mind such considerations as baseboard joints, the position of baseboard framing which may interfere with the location of point motors and the siting of the structures and platforms which you intend to incorporate in your layout.

In station areas and their immediate environs, the deep ballasting found on high-speed main lines is not found, so

Laying and ballasting Peco trackwork with loose ballast is a straightforward process. The cork underlay is laid to the centre of the marked line, then, when the glue has dried, the track is pinned to the underlay and finally, when the track has been tested, ballast is applied loose (granite chips in this case) and brushed into the desired position. A mix of PVA glue and water is then applied by means of a dropper.

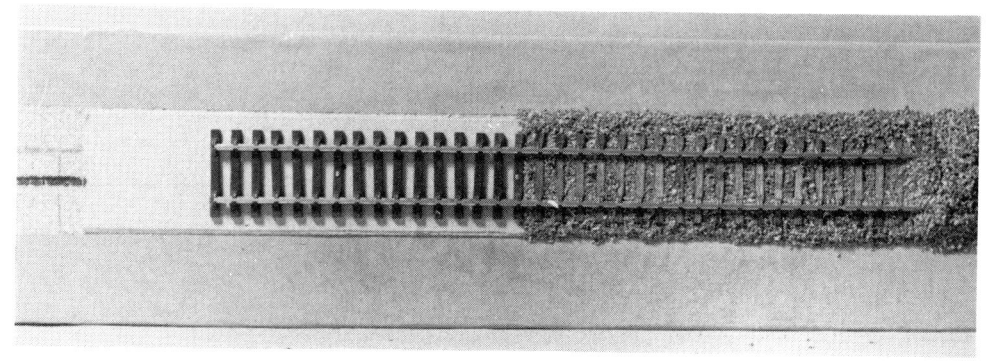

00 gauge 'modern scene' layout _____ 45

Figure 23 Ballasting

Centre line of track marked on baseboard

Cork sheet in half-track strips

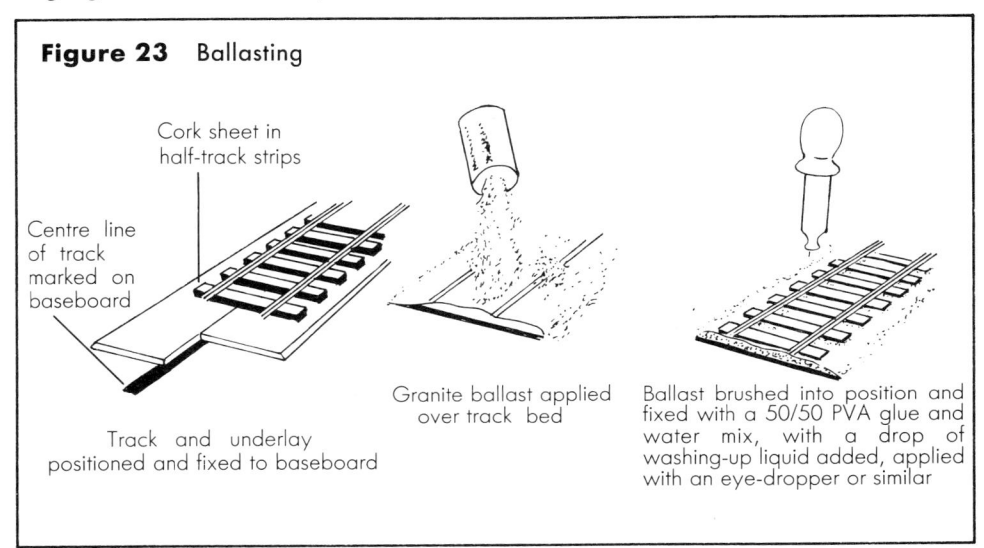

Track and underlay positioned and fixed to baseboard

Granite ballast applied over track bed

Ballast brushed into position and fixed with a 50/50 PVA glue and water mix, with a drop of washing-up liquid added, applied with an eye-dropper or similar

the use of pre-formed foam ballast will not therefore give the most authentic appearance. Track laid on cork sheet, glued to the baseboard and 'loose' ballasted as illustrated in Figure 23 would be more appropriate.

Care needs to be taken in laying the track if good running is to be achieved. It is particularly important to ensure that no kinks occur in the track, that it is flat and level and that the rail joints are smooth and even.

Once the final position of the track is established, it can be cut to length and pinned down to the baseboard. It is better to fasten the pointwork first and lay and adjust the plain track accordingly, rather than vice versa. Ballasting is done once the track is laid, tested and wired.

Wiring

Wiring for the trackwork is fairly simple and a wiring diagram is shown in Figure 24. This shows the minimum necessary to use the layout's full potential, but it could be developed further.

Rather than have three separate leads from each of the three baseboards to the control panel, the wiring could be combined and run cumulatively via multi-pin plugs and sockets to the central baseboard where only one connection to the control panel would then be required (see Figure 25). It is better to use larger plugs and sockets than may be initially required because of the possibility of incorporating electrical

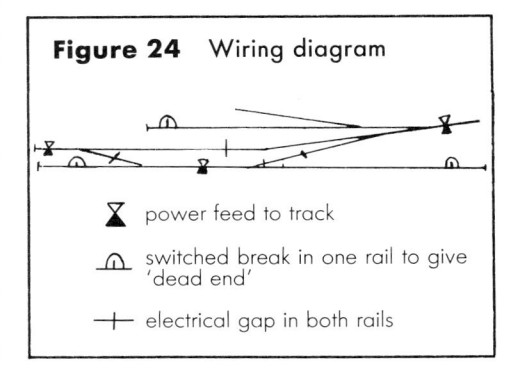

Figure 24 Wiring diagram

⚡ power feed to track

⚊ switched break in one rail to give 'dead end'

┼ electrical gap in both rails

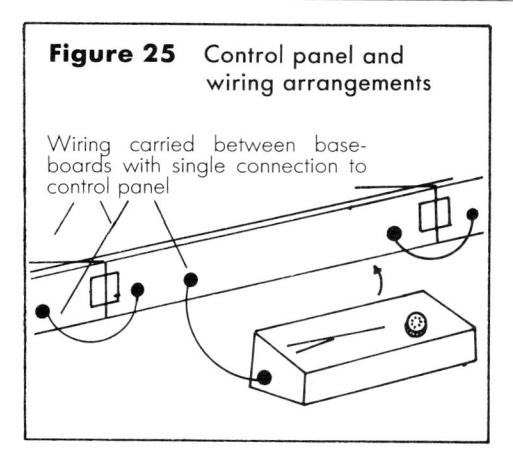

Figure 25 Control panel and wiring arrangements

Wiring carried between baseboards with single connection to control panel

The versatile Peco point motor. Beneath the motor is an accessory switch pack which can be wired to change the polarity of the point and to work colour-light signals and so on. (Peco Ltd)

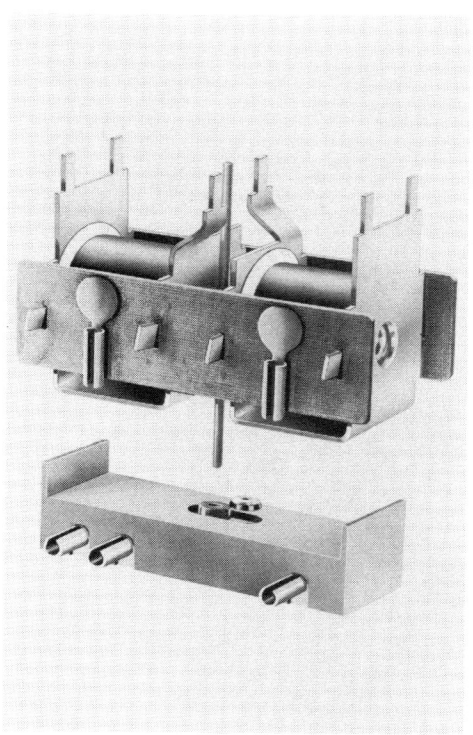

circuits for colour-light signals, ground signals and other accessories at a later date. No wiring diagram has been included for these, since their inclusion is obviously a matter of choice. Suffice to say, however, that the instructions included with these products should be followed carefully or permanent damage may result. The wiring from the baseboard in which they are fixed can be fed through the multi-pin plug and socket circuits back to the control panel where, of course, switches and control equipment should be installed.

Scenery

The basic scenic development centres on the *retaining walls* surrounding the station itself and the two road overbridges. These walls can be easily made from scraps of timber and plywood as shown in Figure 26. They can be as detailed or abstract as you wish but will certainly need covering by a brick or stone paper to represent your desired finish, the most common perhaps being 'engineers' blue brick'. There are also some excellent embossed stone cards available in colours which will match most of the English regional variations. Arches can be incorporated into the walls as illustrated, with capping stones made from balsa wood or modelling clay such as Das, and painted appropriately — sooty black seems the most common colouring!

As far as the *bridges* are concerned it is likely that they would be of cast iron, with stone, brick or even steel pillar supports. Some suggestions for construction are shown in Figure 27. A range of commercially-produced plastic girders and bridge sections is available from Peco and a number of

00 gauge 'modern scene' layout

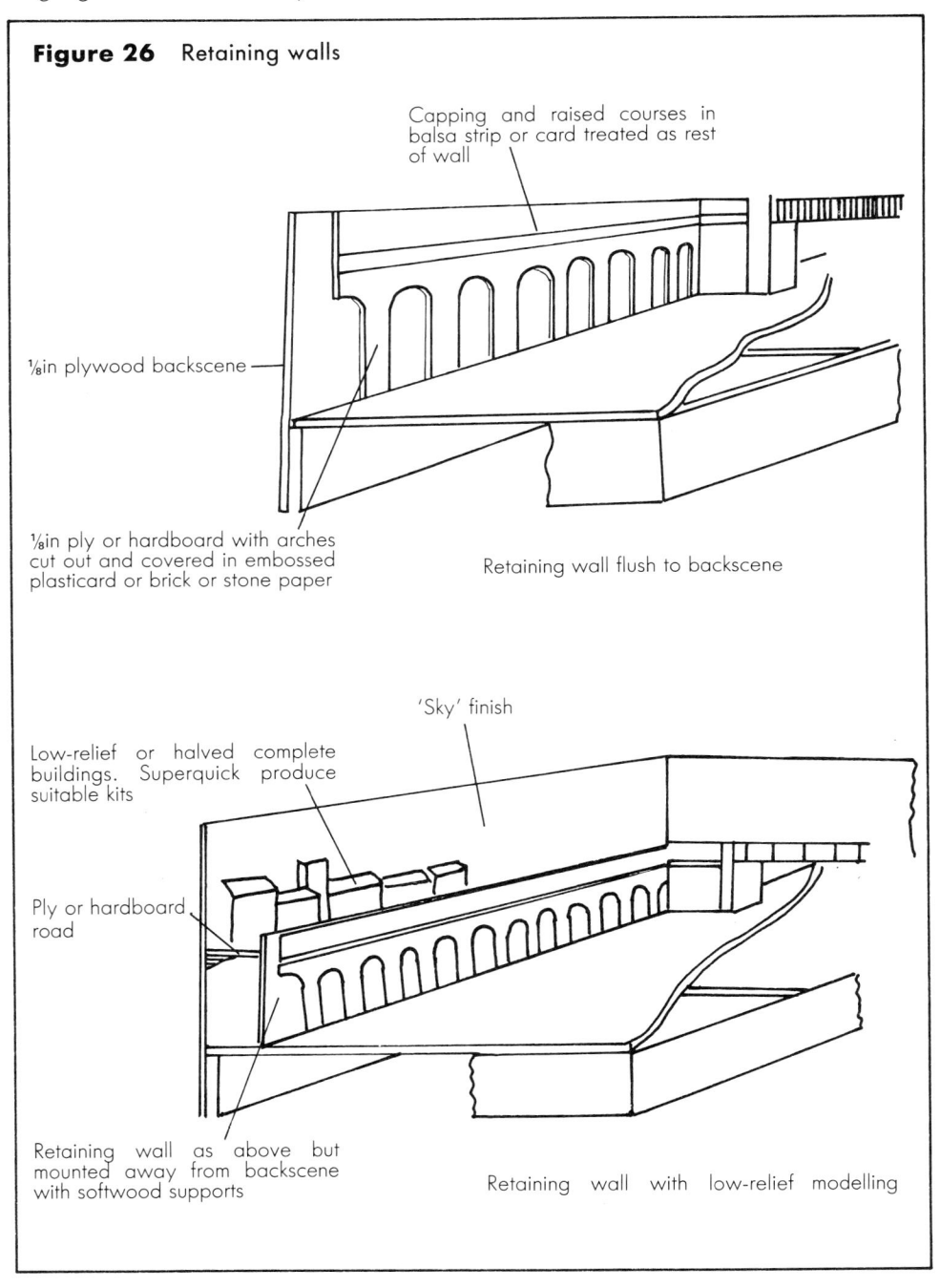

Figure 26 Retaining walls

Capping and raised courses in balsa strip or card treated as rest of wall

⅛in plywood backscene

⅛in ply or hardboard with arches cut out and covered in embossed plasticard or brick or stone paper

Retaining wall flush to backscene

'Sky' finish

Low-relief or halved complete buildings. Superquick produce suitable kits

Ply or hardboard road

Retaining wall as above but mounted away from backscene with softwood supports

Retaining wall with low-relief modelling

Figure 27 Bridge details

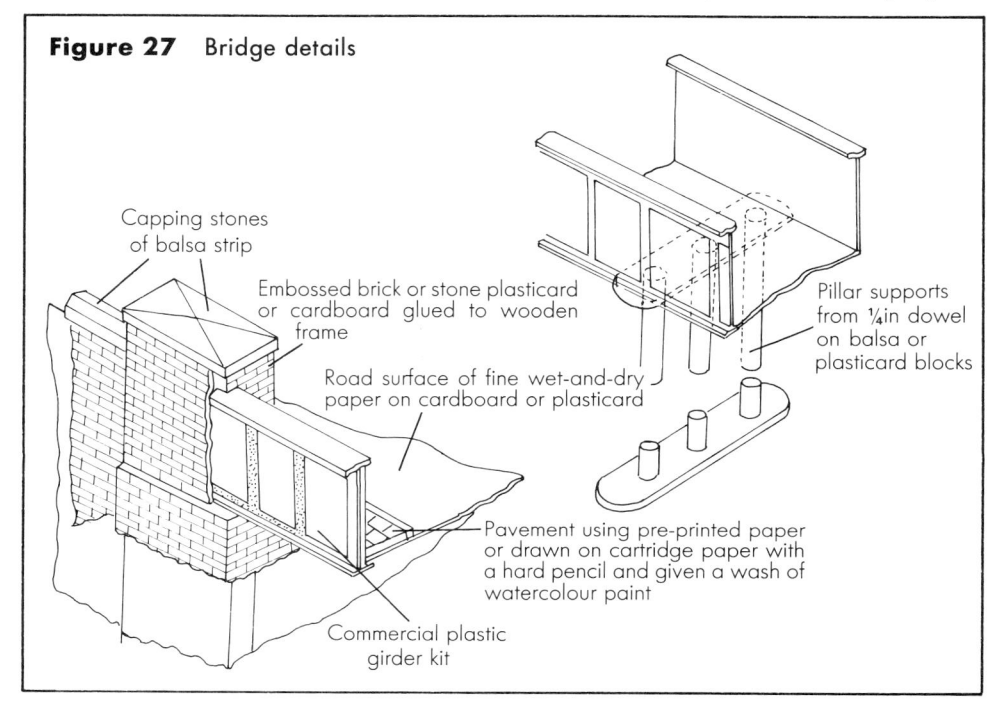

Capping stones of balsa strip

Embossed brick or stone plasticard or cardboard glued to wooden frame

Road surface of fine wet-and-dry paper on cardboard or plasticard

Pillar supports from ¼in dowel on balsa or plasticard blocks

Pavement using pre-printed paper or drawn on cartridge paper with a hard pencil and given a wash of watercolour paint

Commercial plastic girder kit

Whilst the supports of the bridge in this picture are stone, it nevertheless shows the construction of the type of bridge required for this layout. Figure 27 shows the detailed construction.

European plastic kit manufacturers. Alternatively, several different girder bridge kits are available and could be used as an alternative for the road bridge near the fiddle yard.

The road bridge at the left-hand end of the baseboard carries the *station building* which covers several inches of the track; the general construction arrangements are shown in Figure 28. The station building itself could be made from an improved and adapted proprietary building or kit as here, and, if the continental plastic kits are considered, the choice is extensive. However, if a continental kit is chosen, even the more British-looking ones will need some modification to be acceptable. The arrangement shown uses the Hornby building; whilst quite expensive, it can often be picked up quite

00 gauge 'modern scene' layout 49

cheaply in a dealer's second-hand box, although it may have suffered at the hands of 'junior'. However, work would need to be done on it in any case to make a fully-detailed model.

A *flight of steps* for getting passengers down to platform level is required and in reality the original arrangements would probably have been quite elaborate, involving a canopy over the platform. However, in our 'modernised' layout a simple arrangement adapted from part of a cheap footbridge kit and given a rudimentary roof will suffice, as shown in Figure 29.

It is assumed that modernisation has resulted in the removal of the platform canopy and the Victorian platform waiting rooms, and their replacement by a modern glass and plastic *shelter*.

These two plastic bridge sections could be used to advantage in the construction of this layout. There are many other variations of bridge work available from several manufacturers, including Peco and Faller. (Peco Ltd)

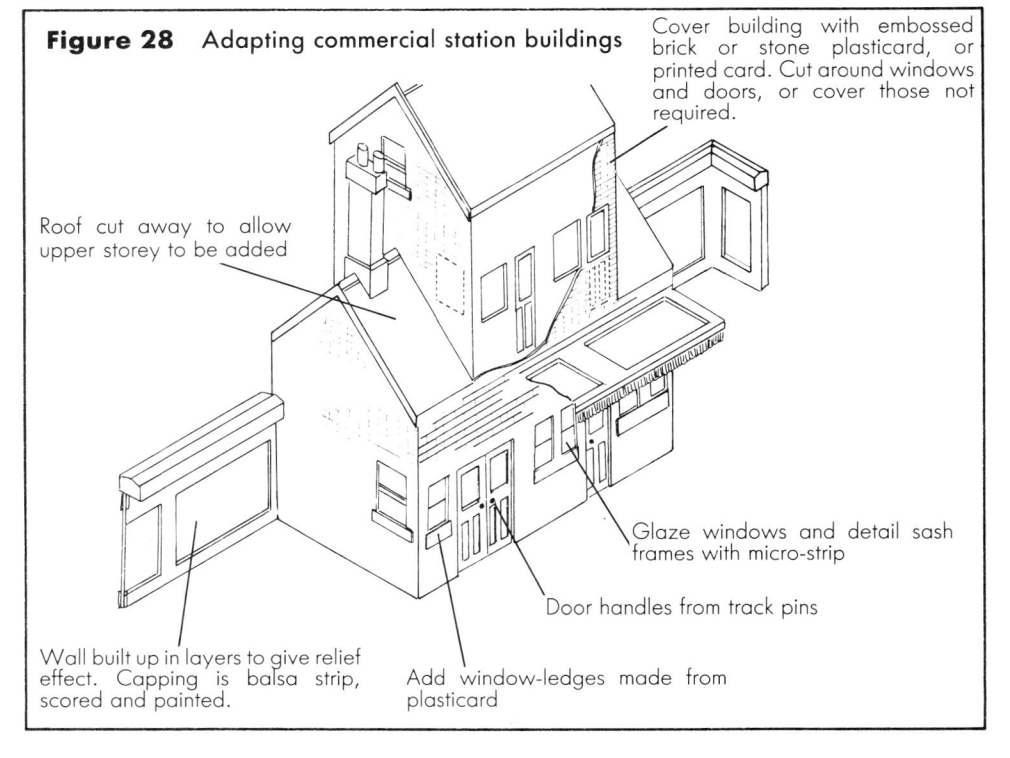

Figure 28 Adapting commercial station buildings

Cover building with embossed brick or stone plasticard, or printed card. Cut around windows and doors, or cover those not required.

Roof cut away to allow upper storey to be added

Glaze windows and detail sash frames with micro-strip

Door handles from track pins

Wall built up in layers to give relief effect. Capping is balsa strip, scored and painted.

Add window-ledges made from plasticard

Simple model railway layouts

Figure 29 Bridge details and platform stairs

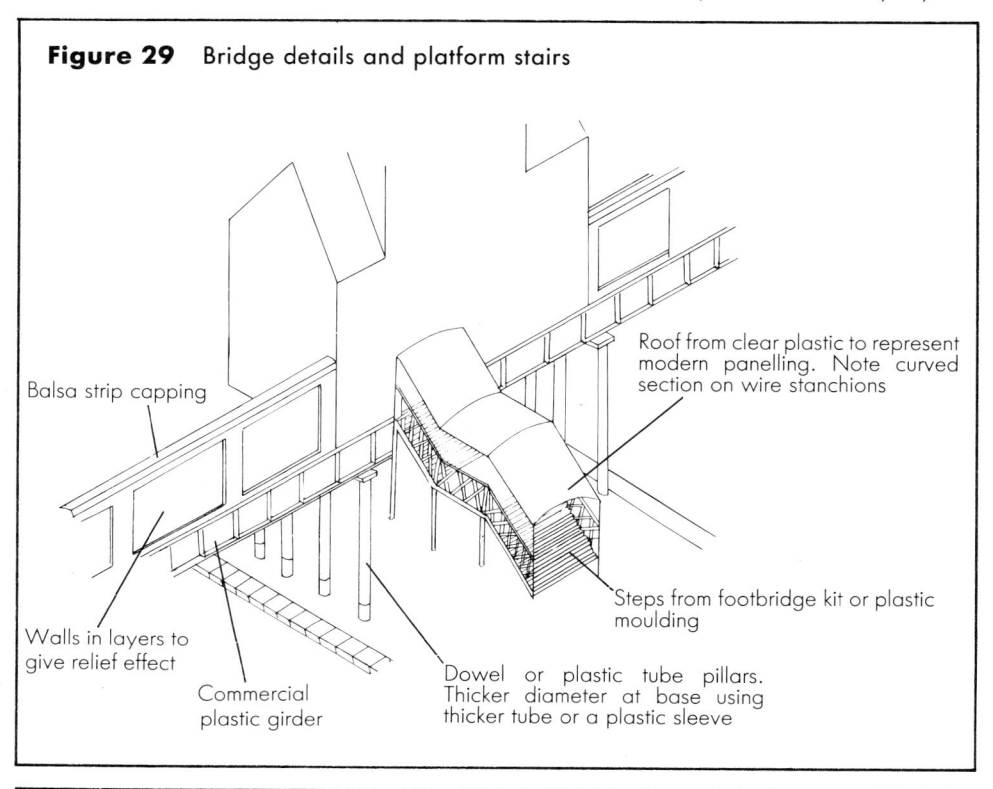

Roof from clear plastic to represent modern panelling. Note curved section on wire stanchions

Balsa strip capping

Steps from footbridge kit or plastic moulding

Walls in layers to give relief effect

Commercial plastic girder

Dowel or plastic tube pillars. Thicker diameter at base using thicker tube or a plastic sleeve

Figure 30 Platform construction

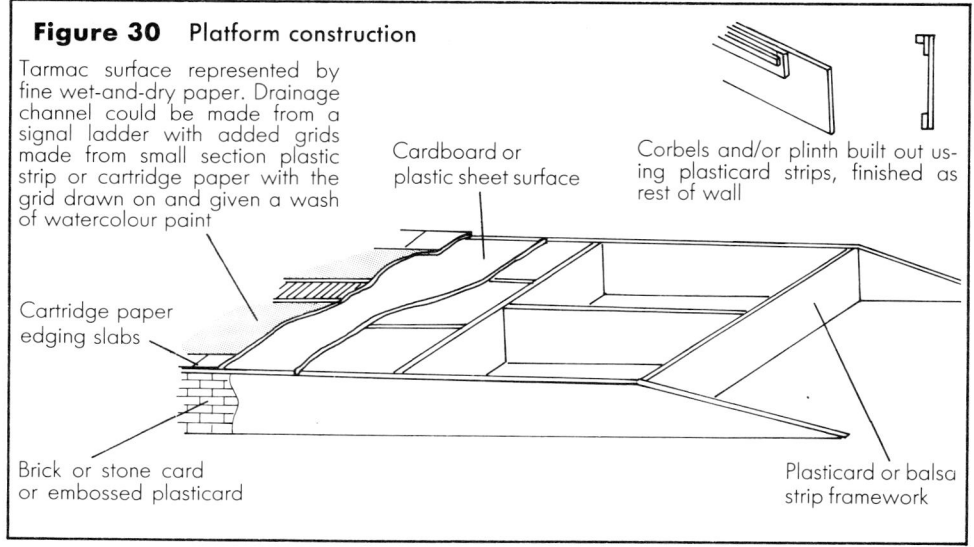

Tarmac surface represented by fine wet-and-dry paper. Drainage channel could be made from a signal ladder with added grids made from small section plastic strip or cartridge paper with the grid drawn on and given a wash of watercolour paint

Cardboard or plastic sheet surface

Corbels and/or plinth built out using plasticard strips, finished as rest of wall

Cartridge paper edging slabs

Brick or stone card or embossed plasticard

Plasticard or balsa strip framework

Peco produce a simple kit for a suitable building. Alternatively, one could easily be made from plasticard and clear plastic sheet, since they are, in their most basic form, often only a glass box with a roof and, if you are lucky, a bench or two inside.

The *platform* itself is easily constructed as shown in Figure 30. The surface would probably be tarmac with a shallow drainage gully down the centre. The platform edging would be large whitish-grey concrete slabs and in all probability the original brick foundation would remain.

Modern working lights are available from a number of manufacturers and could be incorporated with good effect on the platform. Wiring and installation instructions vary from product to product but they are all easy to install. Wiring back to the control panel is via the multi-pin plugs and sockets.

Lots of interest and atmosphere can be created around the old *goods yard* and if, as suggested, this is a local coal distribution depot then some form of transferring and stacking the coal, such as a JCB or crane, a conveyor belt and coal 'bins', probably made from old sleepers, will be required (see Figure 31). The vehicles can be found from amongst the smaller die-cast toys available which, with judicious painting and dirtying, are transformed from toys into models suitable for a model railway layout. Don't forget a high capacity lorry or two for loading.

A *signal box* could be added, and here the choice is yours. If you have a specific location in mind, you will no doubt require a signal box of a particular pre-grouping company, for example the Prototype ex-LNWR signal box kit or the Dapol ex-Midland box. The box would probably be a bit run down with faded paintwork, recreated by 'dry brushing' grey over the painted wooden parts. (Dry brushing is a simple technique where a paintbrush is

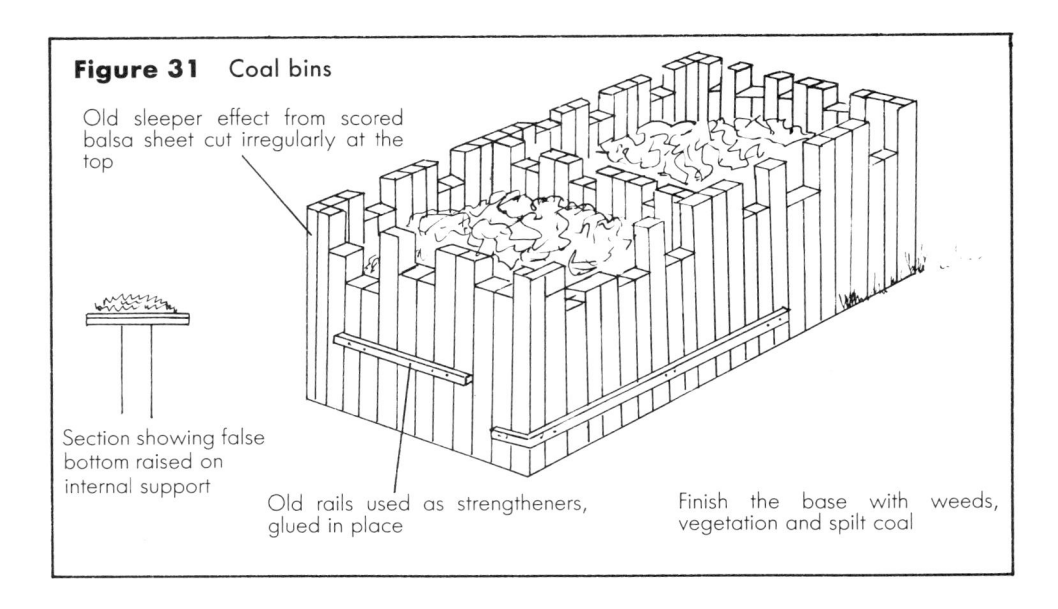

Figure 31 Coal bins

Old sleeper effect from scored balsa sheet cut irregularly at the top

Section showing false bottom raised on internal support

Old rails used as strengtheners, glued in place

Finish the base with weeds, vegetation and spilt coal

loaded with a colour, then squeezed in a paint-rag until it is almost dry; when it is brushed over the area to be treated, it leaves only faint traces of colour.) Add rusted buckets, barrels, old sleepers and, of course the signalman's bike! Platelayers' tools, chairs and a whole range of detailing parts are available from model railway shops; there are even electrical junction boxes and concrete cable-ways for power circuits along the trackside. Close observation of the real railway scene, either 'live' or from the many photographic albums available, is sure to give you countless ideas.

The *mail order warehouse* should ideally be built to represent a specific type or design of building and it would be a simple task to do this from scratch using card covered with brick paper, or one of the continental plastic kits such as the Heljan Brewery which could be adapted to give a reasonable impression of a medium-sized railway company goods warehouse. The building is only modelled in half relief to allow for the fiddle yard behind. An easy-to-construct method of building this feature is illustrated in Figure 32. Modern offices could be provided as a later addition at the end of the old railway building.

Locomotives and rolling-stock

Some examples of suitable rolling-stock and locomotives for the layout have already been mentioned. Happily, there are regular additions to the ranges of diesel locomotives and modern coaching and goods stock provided by major manufacturers such as Hornby and Lima, and kit manufacturers such as MTK. It is impossible to generalise on the locomotive and DMU types operating in particular areas at

Peco make a simple card and plastic kit for a modern building. The kit is realistically pre-coloured and can be adapted to represent a variety of typical modern structures or combined in a number of ways to produce larger complexes. The kits could be used on this layout both to provide the platform shelter and the modern ancillary office accommodation next to the mail order warehouse. (Peco Ltd)

00 gauge 'modern scene' layout ——————————————————————————————— 53

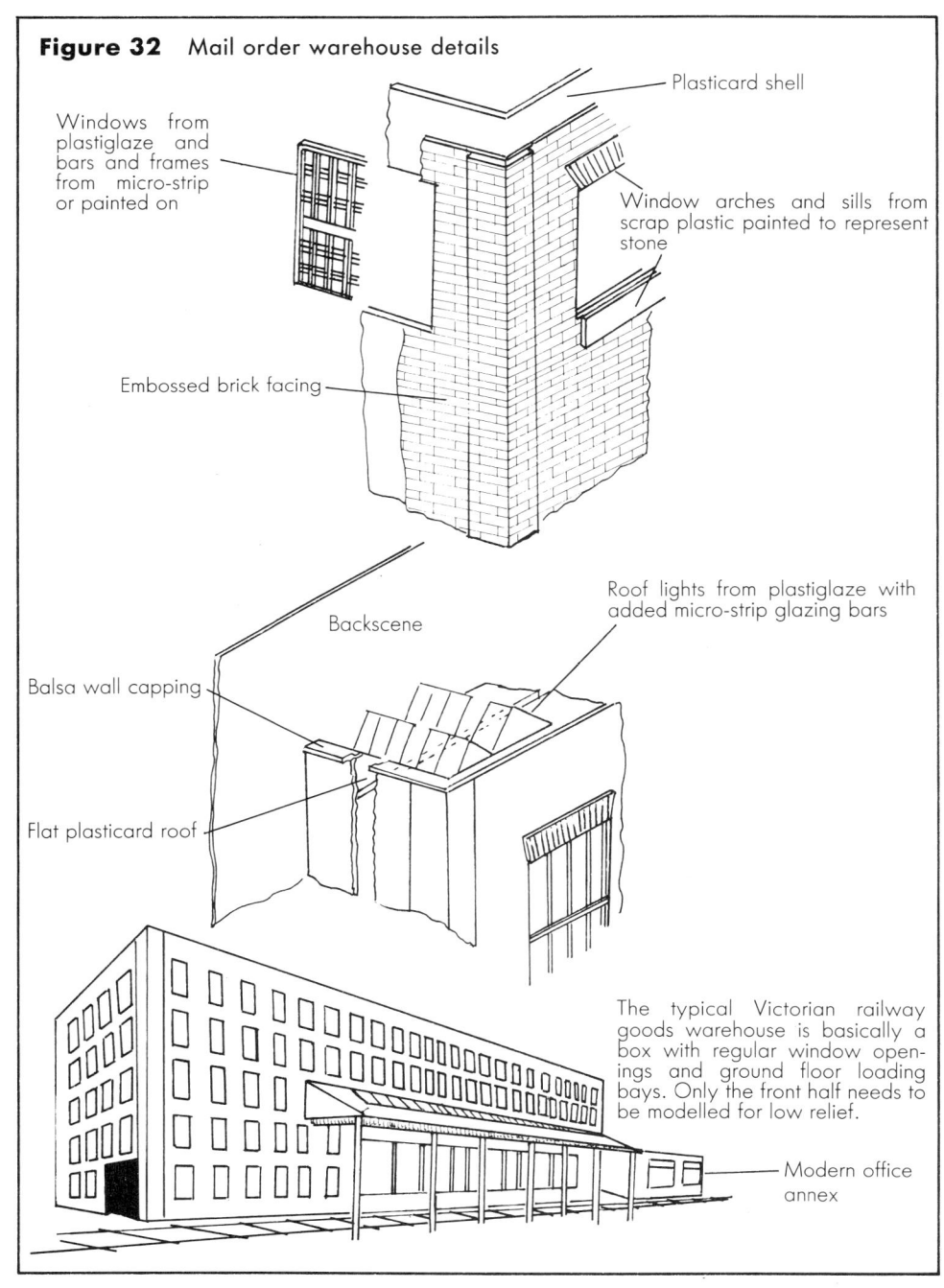

Figure 32 Mail order warehouse details

- Plasticard shell
- Windows from plastiglaze and bars and frames from micro-strip or painted on
- Window arches and sills from scrap plastic painted to represent stone
- Embossed brick facing
- Backscene
- Roof lights from plastiglaze with added micro-strip glazing bars
- Balsa wall capping
- Flat plasticard roof
- Modern office annex

The typical Victorian railway goods warehouse is basically a box with regular window openings and ground floor loading bays. Only the front half needs to be modelled for low relief.

Simple model railway layouts

Above *A lot of interest has been shown recently in modern diesel and electric locomotives and compatible rolling-stock and accessories. Hornby have a number of examples in their catalogue, illustrated here is their Class 25 with a MkII coach.*

Below *The Hornby Class 47, Lima MkI coach and parcels van would be ideal for use on this layout. An industry has now developed providing parts to super-detail readily-available diesel models and stock, ranging from simple flush glazing kits to quite complex conversions to different class variants.*

00 gauge 'modern scene' layout 55

Lima are in the forefront of manufacturers of modern diesel and electric locomotives and rolling-stock and amongst their products are this Class 20 diesel, the DMU and an excellent 'Sea lion' ballast hopper. The DMU has been the subject of numerous articles in the model press in recent years, both for detailing and for conversion to a number of other types.

particular times as classes and types are constantly being reallocated between services and areas as well as being modernised and refurbished. New DMUs such as the class 142 and 150 'sprinters' are now appearing and you will no doubt have your own ideas and favourite classes, especially if you have a particular area and time in mind for the layout. Photographic albums can be helpful here too.

Further references

Diesels in the North West, OPC

Diesels on the Regions Series, OPC

Motive Power Recognition Series, Ian Allan

Railway Magazine

Modern Railways

CHAPTER 4
N gauge main line

This layout project again deals with the modern British railway scene but provides an entirely different concept for the model. If you like to see longer trains of five or six coaches running through the landscape, unless you have a very large room at your disposal to turn your ambition into reality, you may have to look at the smaller gauges and, in particular, N gauge.

There is now an extensive range of British locomotives and rolling-stock of good quality and running ability available, and model diesel locomotives seem particularly effective. Equally important, there is a rapidly-growing range of scenic detail accessories and a well-established range of trackwork.

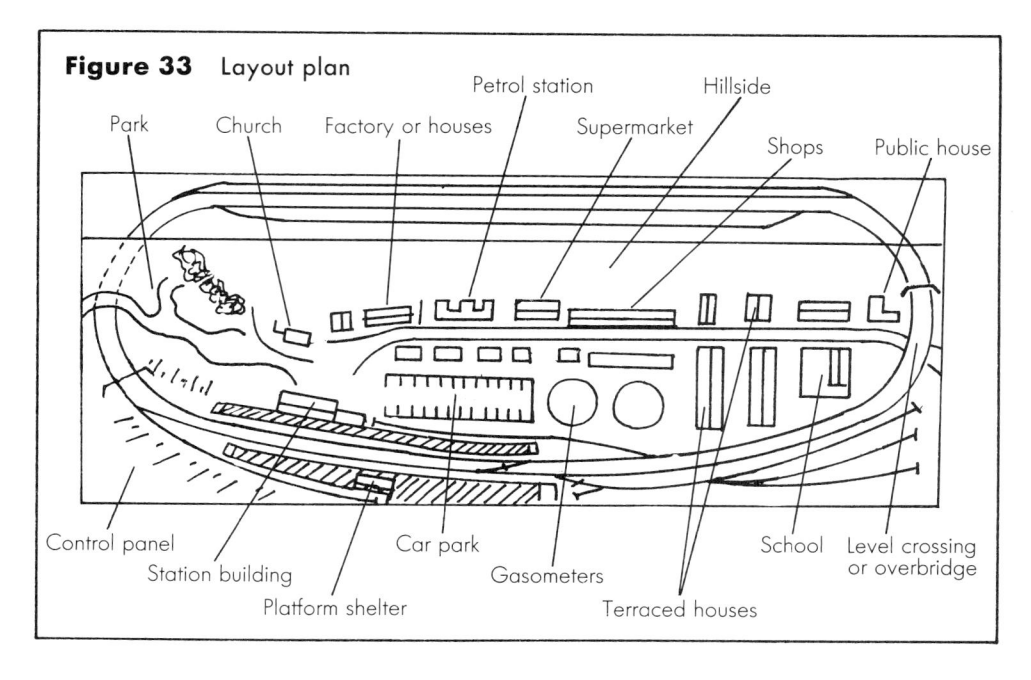

Figure 33 Layout plan

N gauge main line

The commercial flexitracks are often provided with compatible foam ballast underlay, and in N gauge it looks particularly effective (Peco Ltd)

The plan

The plan illustrated in Figure 33 is for a layout representing a secondary main line, and occupies a space of 6ft 6in × 2ft 6in. It provides for two bay roads behind the through platforms on each of the main lines, and sidings to store goods, empty stock, ballast and engineering wagons.

Baseboard construction

The size of the layout is dictated by the size of the baseboard material, which comes almost ready-made from your local builders' merchant; 6ft 6in × 2ft 6in has been chosen as it is one of the range of standard sizes. The baseboard is, in fact, a standard hollow interior door; in first quality these are quite expensive, but seconds would be ideal for the purposes of a model railway baseboard, providing they are not too badly damaged. They are readily available from the builder's merchant for a fraction of the cost of a first quality door. It may seem strange to suggest a door as a model railway baseboard but it does have a number of advantages. Firstly, it is square, stable and the type suggested is light in weight for its overall size. Secondly, whilst it is a large single unit, it can be carried through a normal doorway and up and around staircases; in other words, all the places that a door intended for more conventional use would need to go. Lastly, it is a cheap source of baseboard and requires no framework.

Because of the purpose for which it is designed, the baseboard could be hinged to a bracket on a wall to make a layout which could be folded away when not in use; because it is finished on both sides, the underside would not look as unsightly as a conventional baseboard underframe when it is folded out of use.

Trackwork

For the majority of modellers, the Peco Streamline flexitrack system running on foam ballast underlay would be an acceptable choice for this layout. Pointwork with live frogs is preferable as it gives better slow running, important in starting and slowing trains in

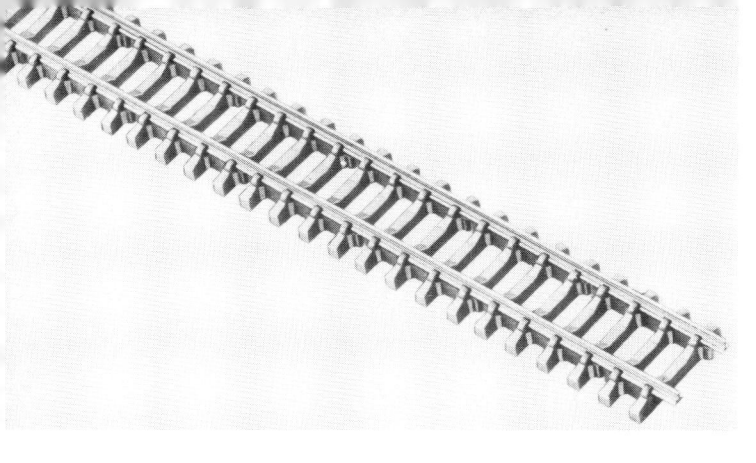

Peco's N gauge concrete-sleepered trackwork is ideal for a modern image layout.

and out of the station and in shunting the sidings.

Normally, if a soft fibreboard surface is used, the track is placed in its foam ballast underlay then curved and laid, held in place by pins through the sleepers. However, the hard surface of the door baseboard makes pushing in the track pins very difficult, if not impossible. An alternative is therefore required. The position of the track is carefully marked on the baseboard and when all adjustments have been made the final position is marked with a thick black felt-tip pen. The area where the track is to lie is then covered with glue (a latex type is suitable) and the underlay and track laid. The pointwork should as usual be positioned before the plain track.

The gentle curve through the station can be easily laid by eye, but, to avoid 'dog-legs' and other inaccuracies which will spoil the running of the trains, use some form of template for the smaller radius curves; a range of curved and straight templates is available for N gauge track. Three are required, 9in, 12in and 15in, and their use is illustrated in the photograph. Start with the smallest radius at the back of the layout near the fiddle yard, increasing the radius towards the front of the baseboard and finally com-

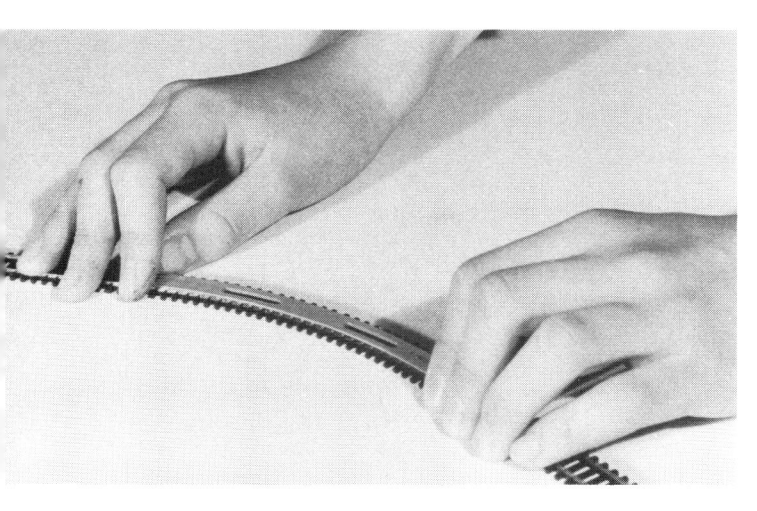

Laying evenly-curved track is quite difficult and, surprisingly enough, laying dead straight track is even more awkward. Thankfully, a range of templates for a variety of radii and for straight track are available from Tracksetta. These are easy to use, and the track can be pinned through the slots in the template whilst held in position.

pleting the circuit by aligning the gentler curves through the station by eye.

It is important when laying the track with the moulded foam ballast that the latter is flat on the baseboard surface and that wrinkles do not occur, particularly on the tight curves. Equally important is care in the joining of the ballast sections; nothing looks worse than yawning gaps or cracks. Good track laying is very important, particularly in the smaller scales, if good running is to be obtained. Care and accuracy are the bywords and will be well rewarded later. Too often, the rush to get something running leads to poorly-built baseboards and badly-laid track which result in poor running and a consequent disillusionment with model railways. The baseboard is ready-made for this layout, therefore only the track-laying is left to cause these problems.

There is nothing hard about laying track or using foam ballast; it is simply a question of patience and not accepting the first attempt if it isn't perfect.

Wiring

The wiring diagram (Figure 34) shows where the necessary power feeds and track breaks are required.

The use of an interior door as a baseboard does throw up one major problem, that of accommodating the wiring and point motors beneath the baseboard. It has to be accepted that both will have to be above the board, but modern developments in electronics come to the rescue so far as the wiring is concerned in the form of the self-adhesive copper electrical tape which is now available. The wiring, instead of being run in conventional wir-

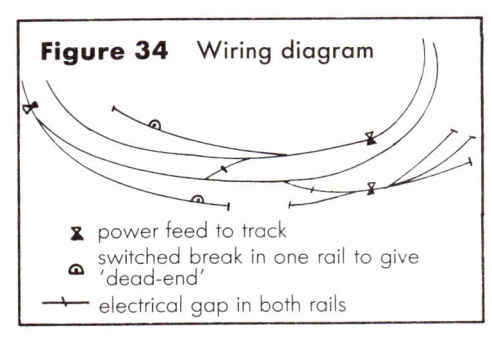

Figure 34 Wiring diagram

✗ power feed to track
⌒ switched break in one rail to give 'dead-end'
—⊢— electrical gap in both rails

ing harness beneath the baseboard, is run on the surface using the tape which can later be be unobtrusively covered with scenery. The tape can be either lifted slightly from the baseboard surface and soldered directly to the rails and the terminals of point motors and accessories, or short connecting 'droppers' or links of conventional wire can be used to bridge the gap. Run the tape neatly along the surface of the baseboard to the corner where the control panel is situated; to minimise subsequent damage, it is a good idea to record where it goes and run it close to, and parallel to, the main running lines. The tape is not too expensive and is available from electronics and radio spares shops and is also often advertised in the model railway press.

It would be possible to wire the layout using conventional wire running along the baseboard surface to the control panel, but to avoid unsightly lumps and bumps in the scenery the very finest wire should be used, available from radio and electronics parts dealers. It should be run either at the very edge of the track to help hide it or in channels carefully scored in the baseboard (or a combination of both). The wire should be covered with broad tape either of the gummed paper or modern thin plastic parcel type.

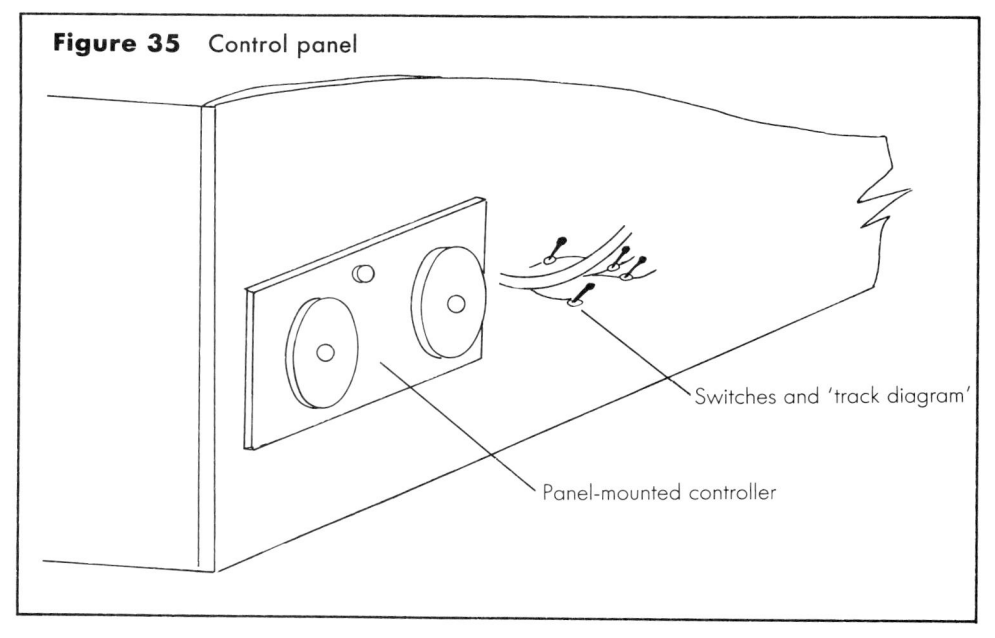

Figure 35 Control panel

Switches and 'track diagram'
Panel-mounted controller

Again, it is wise to record carefully where it runs.

Hiding or disguising the point motors on the surface of the baseboard is quite easy — they can be hidden in signal boxes or other buildings or under scenic features. They do not always have to be immediately adjacent to the points they work and the operating linkage can be extended with piano wire running in a groove chased in the baseboard surface which can then be taped over. However, unless a system of cranks is used, the operation must rely on a straight run of linkage from motor to point.

Be careful when chasing out any paths in the baseboard for either wiring or linkage that you do not go through the skin of the door, as it will only be at most $\frac{1}{8}$in thick. Remember the old adage that you can always take a bit more out but you can't put it back!

The control panel is built into the scenery as shown in Figure 35. It is much easier to make if panel-mounting controllers are used, one for each main line. Miniature toggle switches could be incorporated for the section switches and point control switches, although a stud and probe for the latter would take less space.

Scenery

The basic scenery requirements are for two tunnels through which the railway disappears into the storage sidings at the back, and some high ground near the curve and station at the left-hand end of the baseboard. There is also a station with two platforms, each containing a bay road, and some sidings at the right-hand end of the layout. The rest of the scenic development is a matter of personal choice, urban or rural.

The plan shows an arrangement for

N gauge main line ───────────────────────────────── 61

Right *An alternative to the use of passing contact switches for controlling pointwork is the stud and probe method. The studs are wired to the point motors and installed on the control panel track diagram. The probe is connected to the control panel itself and when any of the studs is touched current flashes to the point motor concerned and activates it. (Peco Ltd)*

Below *This layout illustrates a number of points discussed in this chapter, including the use of commercial bridge and tunnel mouths, the landscape rising to the rear and the bushes and shrubs made from lichen. The station building is from one of the Prototype cardboard kit ranges and is a scale model of an actual GWR building. The backscene finishes off the layout. (Peco Ltd)*

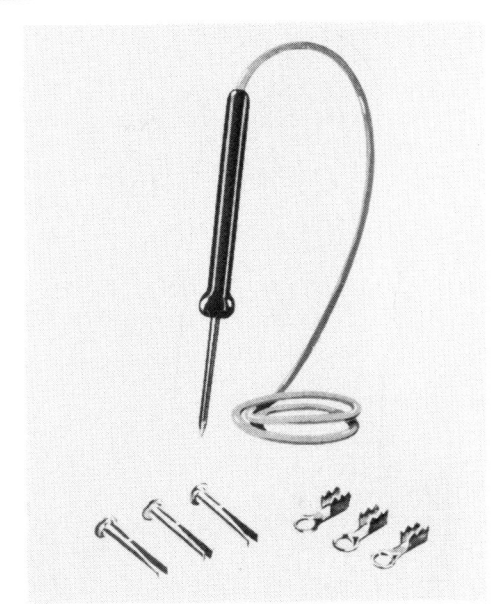

part of a small town, and it is assumed that the rest of the town lies off the baseboard, connected by the road bridge (or alternatively a level crossing). The buildings shown are suggestions only, but this arrangement and degree of urban development is about the minimum commensurate with the level of facilities provided for a railway in the late 1970s and should not tax the ingenuity of the builder or involve so much work as to become tedious.

A start on the scenery can be made once the track is laid and thoroughly tested. The first job is to make the trackwork itself look more realistic. Start by painting it in with a diluted track colour paint, using a spray or a 1½in brush. Most of the model paint manufacturers have this colour in their range. Avoid painting the contacts on the points, and wipe the surface of the rails clean to restore the electrical pick-up. Pick out the sides of the rails with rust colour and add some blue/black gloss paint between the rails in front of the platforms and in the bays to represent the oily drips and leaks from diesel locomotives. Beware of overdoing it — pop down to your local station and have a look at the real thing.

The next task is to make the framework for the landscape, and since this will depend on the effect you wish to achieve, we will concentrate on the effects shown in the plan. A start is made by fastening a length of ⅛in ply standing some 4in higher than the baseboard surface along the whole of the left-hand edge of the board. Another piece is fixed to the front of the baseboard, 4in in height for the first 2ft or so, decreasing to about ½in above baseboard level for the rest. It may be easier to make this from two separate pieces of plywood. Another piece is fastened to the right-hand edge of the baseboard, 4in high for the back 18in, decreasing to ½in high to join the piece of ply on the front edge, thus completing the three sides. These pieces provide the frame of the layout and the chosen landscape features (see Figure 36).

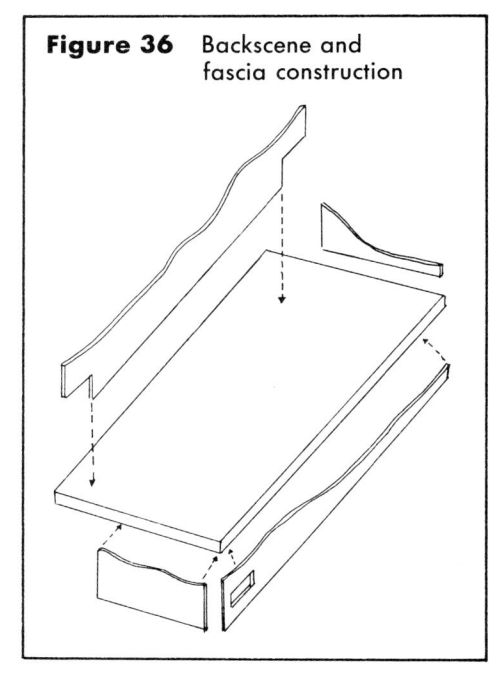

Figure 36 Backscene and fascia construction

The two tunnel mouths can now be positioned, ensuring adequate clearance for the trains to pass through without fouling them. Excellent plastic mouldings are available from Peco, and these include side walls.

The two side frames are linked by a length of ⅛in plywood, cut to allow trains to pass through, and profiled to form the 'skyline' of the landscaped backscene of your choice (see Figure 36). For the purpose of the layout illustrated, a general 'rolling' profile will suffice and is easily achieved even with an ordinary hand-saw.

Attention can now be turned to filling in this frame and making the landscape beyond the town and immediate station area. There are a number of alternative methods of doing this and the two suggested have been well tried and tested.

Firstly, there is the use of small mesh wire netting or chicken wire. This can be bought by the yard from an ironmonger and is cut with side-cutting pliers or shears to fit over and around the tunnel mouths and other features. It is fastened to the framework and baseboard with small wire staples. The netting can be sculpted into the desired shape; for example, the area over the left-hand tunnel is to be a park and therefore the netting base needs to be quite smooth. Conversely, a ragged rock or cliff-type edge and evidence of some ancient water course may be required. To do this, the wire netting needs to be accurately manipulated and bent and twisted to form sharp angles and depressions. Some additional support from scraps of timber may be necessary.

The second method is to use plaster bandage which is put in position over a shaped base created from crumpled newspaper. The newspaper can be removed when the bandage dries, leaving a hard shell which is later treated, coloured and detailed.

Once the basic shape and contours are established, the netting framework is covered in papier mâché (strips of newspaper dipped in wallpaper paste) or with a plaster bandage which is applied, moistened, in strips or squares. When these coverings are dry, a hard shell is formed which can be dressed with the appropriate scenic dressings and detailed with walls, fencing, hedges, trees and rock and soil surfaces.

Vary the shades of green and divide the landscape into fields. Add livestock from the many cast metal and plastic

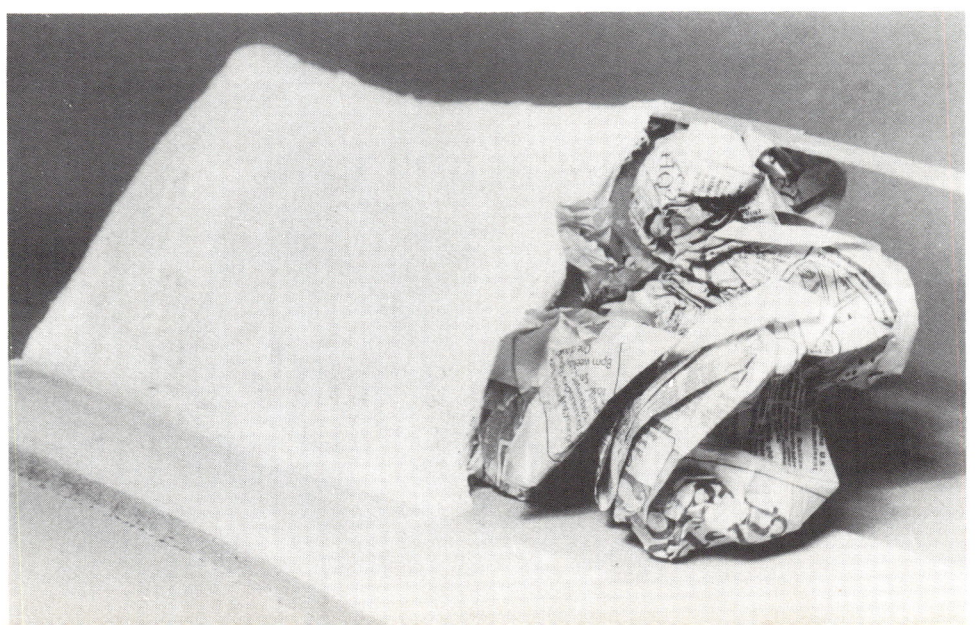

The construction of scenery using a very rudimentary frame, crumpled newspaper and plaster bandage.

types available. Careful painting helps breathe a bit of life into the animals; even the ready-coloured examples benefit from having the shiny plastic toned down and the colour of animals varied. Look at the real thing to get some idea of the variety of colour and in particular the pose and grouping of animals in the landscape. It looks more natural and adds pockets of interest to the layout if scattered groupings of animals are realistically located, perhaps near a farm gate or beneath a tree in a hedge row, rather than spread too evenly over large areas.

When adding colour and texture to the landscape, care should be taken to vary them and avoid large expanses of the same shade. For instance, beside and beneath field gates, change the colour to black/brown earth and create the muddy and rutted earth using a small pool of very runny Polyfilla which can be either coloured at the mixing stage with powder paint or painted with a matt paint when dry. Create the hoof prints of animals by pressing a pin-head repeatedly into the mix when it is almost dry, and make wheel ruts by either running a toy car through the mix as it is drying or draw them with an ice-lolly stick or similar. Fill in some of the depressions in the surface with little pools of gloss varnish to represent puddles.

If you group some cattle by a hedge or under a tree, ensure that the surface is at least earth-coloured to show regular use by the animals as a standing place. You could even provide a field towards the lower level which has been ploughed and this can be represented by coating the area with a coarse, dryish mix of Polyfilla which has already been mixed with a watercolour powder paint of your chosen earth colour. This is then 'furrowed' with a fine comb (Figure 37). Don't forget the areas at the edges and corners where the tractor has turned and left deeper, circular ruts. Add a tractor if desired (there are a number of models available from manufacturers such as Langley) and perhaps a scarecrow (easily made from a plastic figure with the legs removed and impaled on a painted pin pushed into the baseboard).

The town area presents a different challenge and perhaps the first area to be tackled should be the *park* above the tunnel on the left of the baseboard.

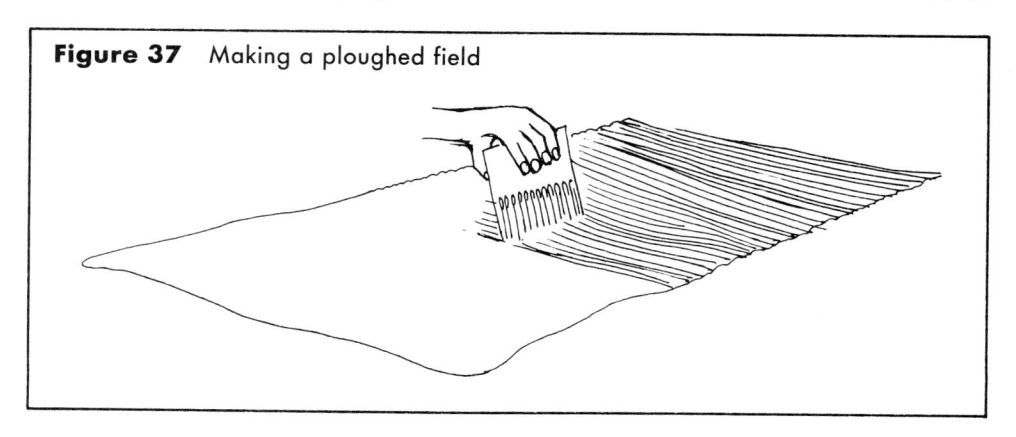

Figure 37 Making a ploughed field

Figure 38 Lawns and paths

Stripe effect lawn: cover alternate strips with masking tape, apply scatter material then remove and use different shades of grass scatter material for the other strips

masking tape   grass

Flower bed cut out of lawn

Scenic dressing

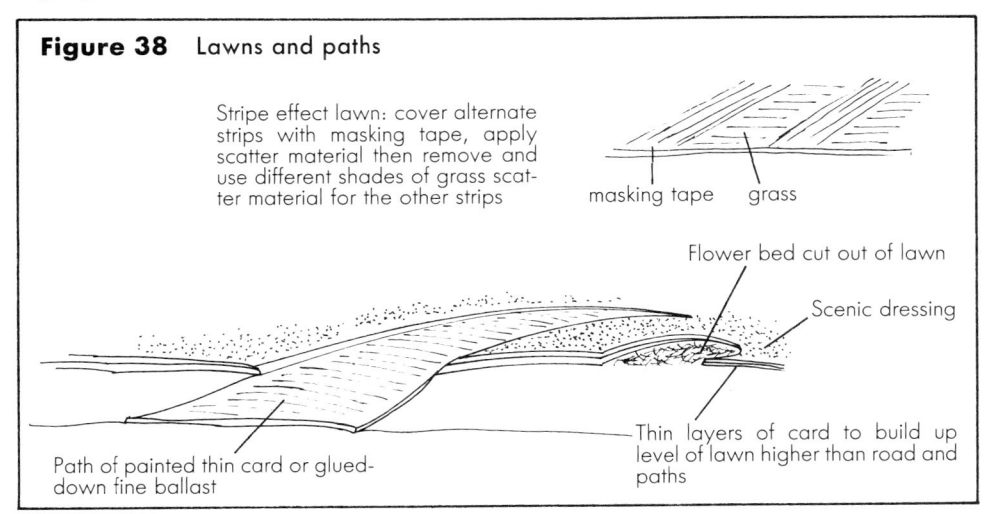

Path of painted thin card or glued-down fine ballast

Thin layers of card to build up level of lawn higher than road and paths

A small flat area could be provided near the gate with simple children's playground equipment mounted in a tarmac surface represented by a small oblong of fine wet-and-dry paper. The boundary of the park could be defined with some of the beautiful etched spear-point iron railings available from the Scale Link range and park gates of the ornamental iron type so favoured by the Victorians and Edwardians are also available. The inside of these railings could be lined with large bushes; lichen painted and coated in dyed sawdust could be used to represent rhododendrons and the other types of plant often found in parks. The 'bed' in which these are 'planted' should be coloured to represent earth.

The paths can be made from fine wet-and-dry paper to represent tarmac surfacing. Lawned areas, slightly higher than the path surface, can easily be made by using thin card cut to the desired shape, painted earth colour and then coated in the finest scenic dressing you can find. You can even represent a striped lawn effect by using two shades of green scatter material, one slightly darker than the other, and coating them in alternate strips, covering the adjacent one with masking tape 2 or 3mm wide.

Cut out shapes in the lawn, fill them with earth mix as used for the ploughed field and add clumps of scatter material dotted with bright red, yellow and blue and you have flower beds. Figure 38 illustrates these methods.

Disguise the transition from park to open country with a line of bushes smaller than those near the iron railings to help create an illusion of distance. Add some benches made from station platform seats and do not forget some people sitting or walking in the park and children in the playground. Lots of well-detailed and painted figures are available from Preiser, Peco and other manufacturers.

The *town* itself can be modelled very effectively by mixing the types of model buildings used to represent the typical mix of property found in to-

Figure 39 Typical range of building elevations

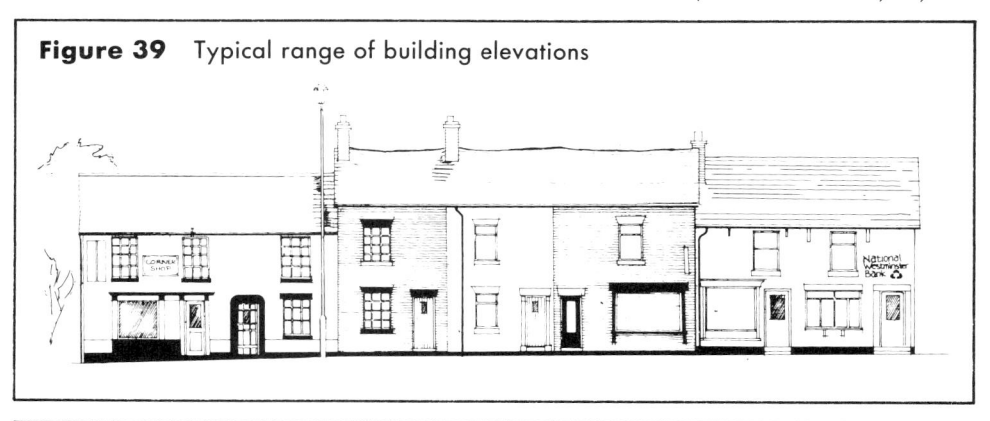

Figure 40 Building from scratch

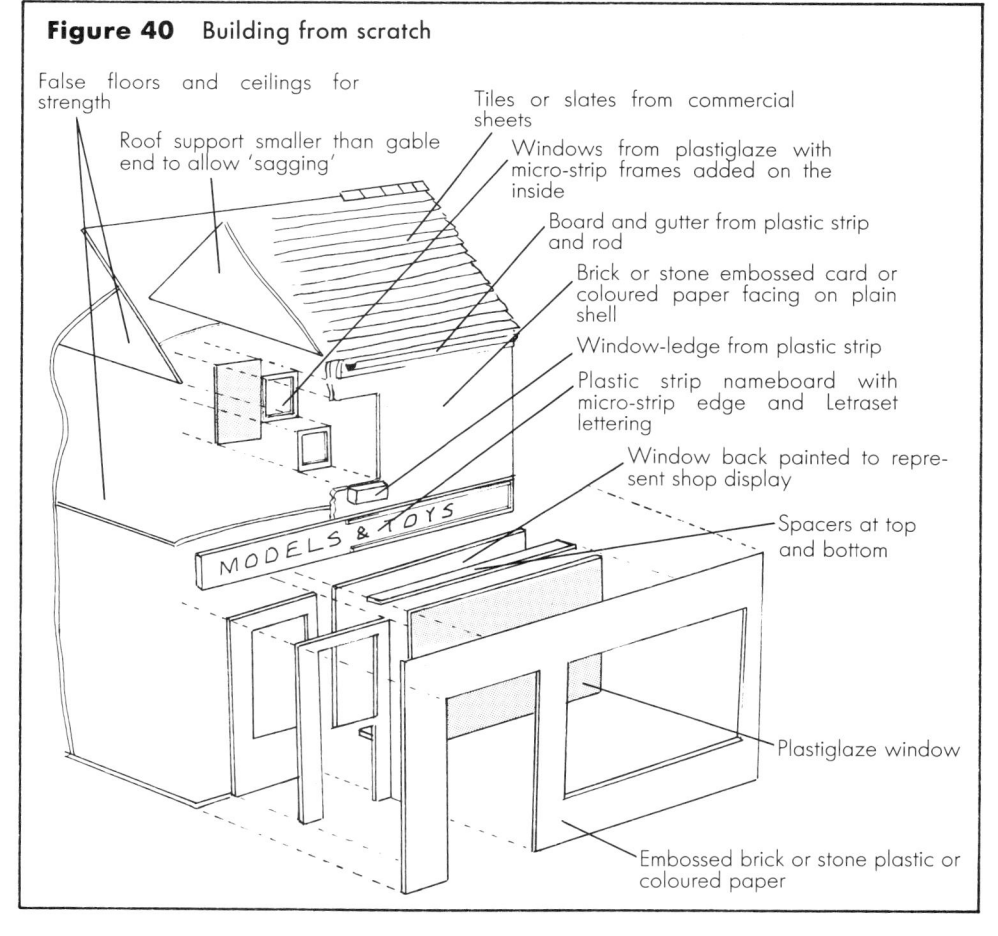

False floors and ceilings for strength

Roof support smaller than gable end to allow 'sagging'

Tiles or slates from commercial sheets

Windows from plastiglaze with micro-strip frames added on the inside

Board and gutter from plastic strip and rod

Brick or stone embossed card or coloured paper facing on plain shell

Window-ledge from plastic strip

Plastic strip nameboard with micro-strip edge and Letraset lettering

Window back painted to represent shop display

Spacers at top and bottom

Plastiglaze window

Embossed brick or stone plastic or coloured paper

N gauge main line

day's towns, brick and concrete, old and modern, side by side.

Some fairly typical front elevations of buildings to be found 'just off the town centre' are shown in Figure 39; it would be quite a simple matter to construct buildings like these and Figure 40 shows how it might be done. Alternatively, Graham Farish produce quite effective traditional terraced houses and shops and these could be used. They are printed but have no relief detail and are therefore perhaps better located away from the foreground. Alternatively, they could be detailed and some relief added such as door surrounds and stone window bottoms, from card or plasticard, and guttering and drain-pipes from plastic rod (Figure 41). What about the odd TV aerial made from fuse wire?

More modern buildings are perhaps easiest to represent by utilising the plastic kits available. A modern supermarket looks much the same in Germany as in the UK, providing the

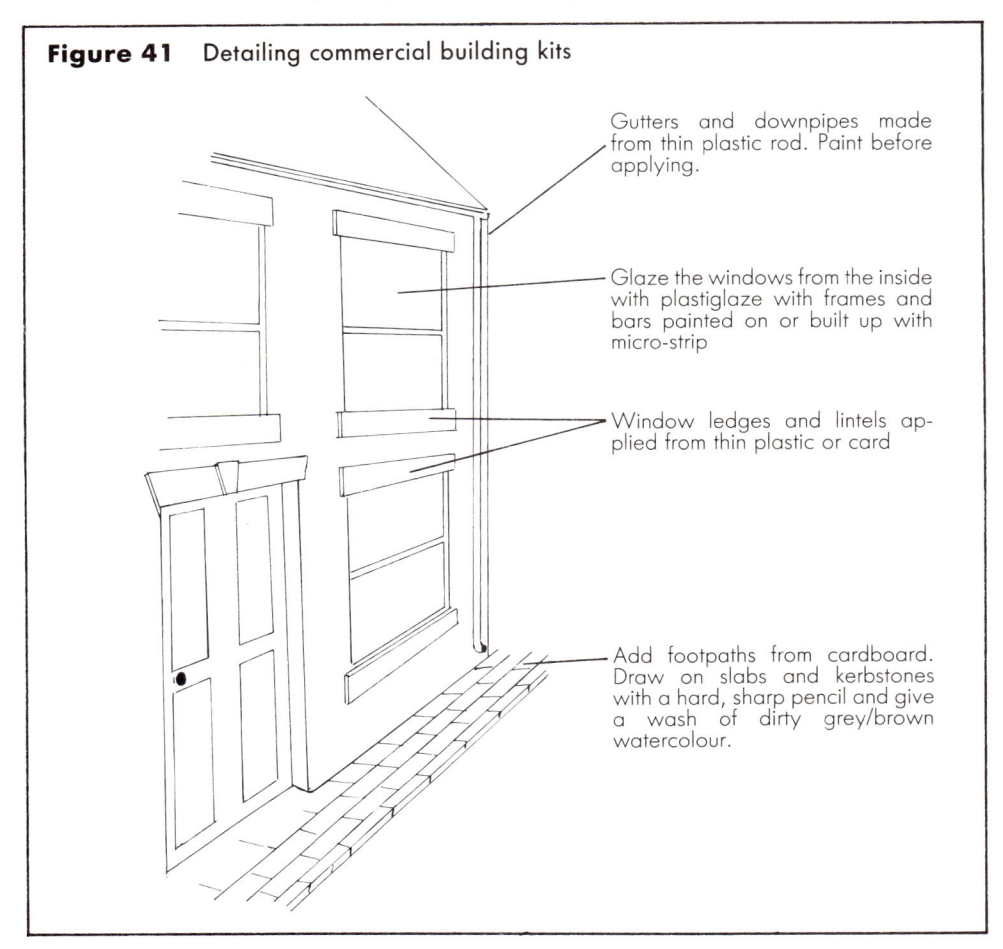

Figure 41 Detailing commercial building kits

Gutters and downpipes made from thin plastic rod. Paint before applying.

Glaze the windows from the inside with plastiglaze with frames and bars painted on or built up with micro-strip

Window ledges and lintels applied from thin plastic or card

Add footpaths from cardboard. Draw on slabs and kerbstones with a hard, sharp pencil and give a wash of dirty grey/brown watercolour.

names are changed, and the range of foreign plastic building kits such as those of Faller, Pola and Heljan can provide a variety of suitable buildings, from supermarkets to filling stations, to mix in with the traditional styles.

Roads are best represented by fine wet-and-dry paper, on which various road markings can be added using either Letraset white lines and letters or a paint stick. Grey is better than white in this scale as the white looks too stark. Footpaths can be added from thin card, preferably white, on which the joints between paving-stones and kerbs can be drawn with a hard pencil. They can then be coloured with a brown/grey wash of watercolour paint.

Traffic lights, road-signs, lamp-posts, cars, buses and lorries are all available, painted or unpainted, and together with the people add a final touch which can bring the layout to life.

The area of the layout to the right of the station, between the town and the railway line, is an area which would have been the goods yard in earlier days. The most likely use for the site in later years would be a car-park, and, to add some relief and further interest, at the other end of this area a gasometer can be constructed. There are excellent plastic kits available for these structures.

The *car-park* is easily made from wet-and-dry paper with the markings added in a pale grey or cream. The area can be fenced round with mesh fencing from the Scale Link range, posts being made from $1/16$ in square balsa wood or plastic section. The car-park attendant's cabin can be made from either a suitably modified railway hut, repainted in the usual blue, or a modern telephone box for a more recent structure. Barriers are easily made from scraps of plasticard and balsa or can even be poached from inexpensive plastic kits of lifting barrier crossings.

As far as the *station* itself is concerned, it is in every way a conventional structure. Construction of the platform follows the basic principles shown in earlier chapters. Minitrix produce a medium-sized plastic station building which, with the addition of some detail

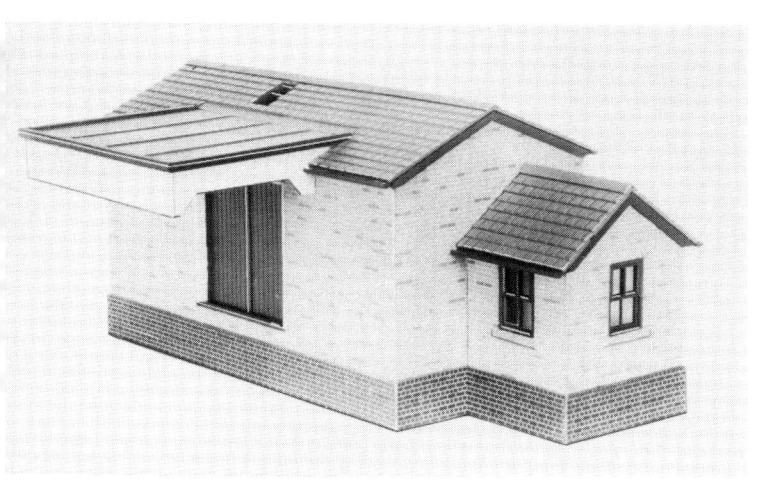

Left *Railway stations which have long since lost their goods facilities often still retain the goods shed, commonly let to a local builder as a yard or warehouse. This could be an alternative to the gasometer and, if so, a commercial goods shed kit, such as the Peco one shown here, would make an ideal base.* (Peco Ltd).

Above *N gauge is attracting considerable support from model manufacturers. Here, the excellent Grafar Class 25 is seen in a Ratio station* (Ratio Plastic Models Ltd).

Right *Graham Farish's highly-detailed N gauge Class 37.*

and careful painting, makes an ideal main building for the inner platform. A simple waiting shelter will suffice for the other platform and can easily be made from plasticard. Station fencing, lamps, seats and people complete the scene.

Operation of the layout can be quite varied. There are the up and down main lines on which run the trains stored in the hidden sidings. Alternatively, a DMU service could be operated from bay platform to bay platform. The bays could also be used for loading and unloading parcels or newspaper traffic. The sidings could be used for the storage of empty stock and goods vehicles which could be exchanged with the vehicles kept in the storage yards for use on the main line.

Locomotives and rolling-stock

There is a very wide choice of modern locomotives and rolling stock from manufacturers such as Graham Farish and Peco which keep up to date with livery changes on British Rail. Most of the major diesel classes are represented

Simple model railway layouts

Left *The Grafar Class 20 with traditional steel mineral wagons make an excellent combination for freight trains and are a contrasting alternative to trains of colourful, fully-fitted 100-ton bogie tank and container wagons. This unfitted mineral wagon train (unfitted wagons were grey) would require a brake van at the rear.*

Left *Modern rolling-stock is not a problem in N gauge. Excellent models are readily available, such as this MkI coach, and those from, for example, Graham Farish cover most stages of carriage development, from the ubiquitous MkI to the latest MkIII types, in a variety of liveries.*

Below *Kit building, whilst not as common in N gauge as in the larger scales, is still available to those who enjoy this aspect of the hobby. The range extends from simple clip-together plastic wagons, such as the brake van shown here, to etched and cast body kits for fitting on to commercial chassis.*

N gauge main line　　　　　　　　　　　　　　　　　　　　　　　　71

A few examples from the extensive range of wagons available for a diesel era layout. (Peco Ltd)

in the Graham Farish range, as are the various types of post-war British Rail coaches from the MK1 stock of the 1950s to the latest MK111 coaches in the new 'executive' livery. A DMU is also available in various liveries from the green of the late 1950s and early 1960s to the later blue/grey livery.

There is a tremendous range of high-quality goods rolling-stock, including modern liveries on traditional vehicles and the later four-wheeled tank wagons from Peco. An increasing range of kits for specialist vehicles and DMUs is becoming available and keeping a watchful eye on advertisements in the model railway press is a good way of keeping abreast of new products in this fast-developing area of the model railway hobby.

Further references

Diesels Nationwide (several volumes), OPC
Freight Trains of BR, OPC

CHAPTER 5
An American 'shortline'

The term 'shortline' is peculiar to North American railways or, more accurately, 'railroads'. It is, however, a misnomer because shortlines are very often far from being short and, indeed, some are quite long even by the standards of British main lines. The definition of a shortline is actually determined by the revenue it generates, not its length.

There were, and indeed still are, American railroad companies connected to but independent of the major systems, usually standard gauge and built for a specific purpose, perhaps to serve a particular community or industry. Shortline systems vary in their characteristics, size and complexity, and range from a basic line which had perhaps one or two second-hand locomotives, cabooses and the odd vehicle, to those with a reasonably sized stud of locomotives and some stock of their own sometimes operating two or more adjacent lines. Most existed to move freight and whilst in the past some may have operated some kind of passenger service it is unlikely to be the case now.

The plan

Given that the source of traffic is freight, and that US railroad locomotives and stock can be quite sizeable, the plan in Figure 42 has been designed to represent the end of a small shortline

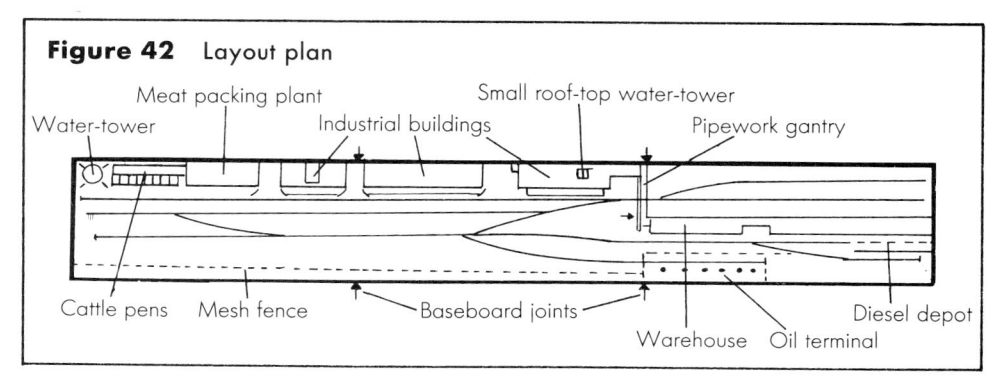

Figure 42 Layout plan

Labels: Meat packing plant, Small roof-top water-tower, Water-tower, Industrial buildings, Pipework gantry, Cattle pens, Mesh fence, Baseboard joints, Warehouse, Oil terminal, Diesel depot

Above *The EMD-SW1 Switcher is an ideal type of locomotive for this layout. The prototype was built over a long period, the earlier examples sporting a large sandbox at the front. The model shown is made in Yugoslavia, sold under the AHM label and, for its cheapness, has quite a remarkable performance, which would shame many more expensive models. The example shown is finished in blue and yellow and decorated for the Santa Fe company. Detailing and conversion kits are available from specialist American retailers for a great many prototypes and a number of articles have appeared in British modelling magazines on detailing and converting these inexpensive models.*

Right *Shunting the 'Pike'.*

which serves an industrial and warehousing complex. It is basically a shunting layout or 'Pike' as the Americans call them. Locomotives will bring in and take away trains from the rest of the rail system. Short trains comprising just a handful of cars will be the norm, but it can be very absorbing and time-consuming distributing locomotives and thus easier to accommodate.

Baseboard construction

The layout has been designed for HO scale (3.5mm to 1ft, 16.5mm track gauge) for which there is an extremely large range of American locomotives, rolling-stock, buildings and accessories available in the UK. It occupies an area of 9ft × 1ft 3in, and the baseboards can be conveniently divided for storage and transport into three separate units of 3ft × 1ft 3in, the smallest practical baseboard size minimizing breaks and joints in scenery and structures. As a more traditional alternative to the method of plywood baseboard construction featured in earlier chapters this is framed with 2in × 1in softwood with a chipboard surface. Good-quality, well-seasoned softwood is virtually unobtainable these days, but the local newspaper will often reveal advertisements for sawn and planed reclaimed timber which has been recovered usually from demolition or other work and is consequently well seasoned. It is also usually as cheap as if not cheaper than the unseasoned variety. If, however, it is unavailable, unseasoned timber usually presents no problems; the author has constructed baseboards from new timber obtained from the local woodyard which have been in use in a variety of conditions for the last four years without warping or twisting.

The advantages, if there really are advantages to this type of traditional construction, is that it may be slightly cheaper and the materials may be more easily obtained and perhaps more easily prepared by the average DIY man. If you prefer, the plywood-framed

An alternative or perhaps a supplement to buildings is a simple storage compound, fenced off with the same materials used for the security fence. In this example, lumber made from balsa has been stacked.

the vehicles to their correct locations, stock-cars to the cattle pens and the oil-tanks to the oil terminal. A fair degree of varied operating can take place and some form of automatic coupling such as the Kadee type would be ideal.

A small depot and repair shop has also been included to service diesels and hidden fiddle yard facilities represent the rest of the line.

A diesel depot has been suggested because there are some excellent American diesel locomotive models available, they don't need turning to avoid tender first running and they are generally somewhat shorter than steam

An American 'shortline' _____ 75

Figure 43 Baseboard construction

½in chipboard

2in × 1in softwood frame

Half-lap

Butt

Centre joint alternatives

Loose-pin hinges to join baseboards

Corner joint, screwed and glued

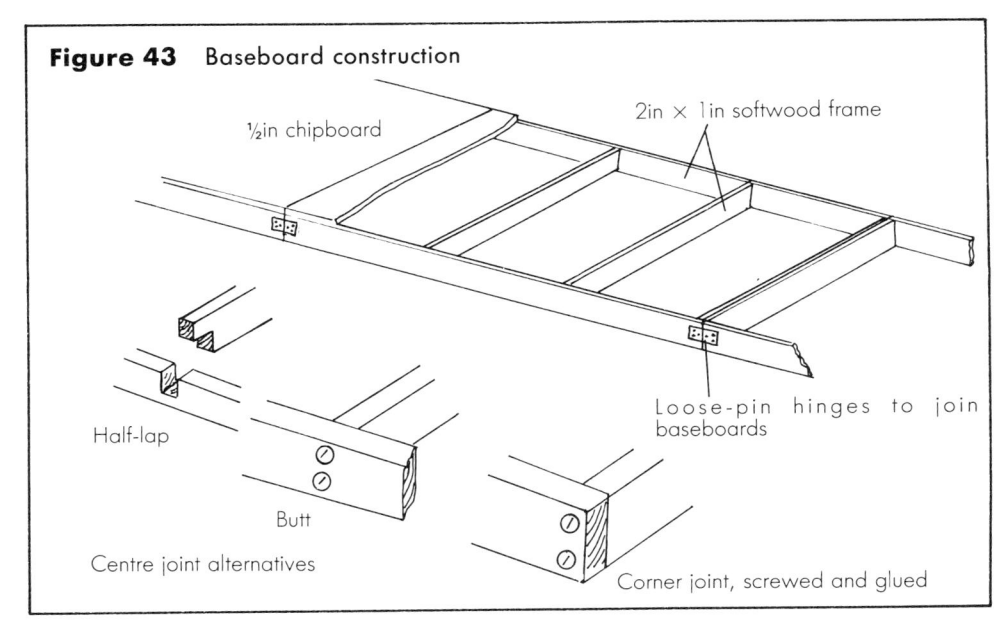

baseboards as described in earlier chapters could, of course, be used and would be quite suitable for this layout.

Construction of the baseboards is a simple exercise and is shown in Figure 43. Basically, there is a softwood grid framework with cross pieces in the middle and at both ends. If you are handy at carpentry, a basic pinned and glued half-lap joint can be used to join the cross pieces and side members. Alternatively, butt joints (saving any cutting of the timber other than to length) with the joint screwed and glued would suffice. Brass screws are preferable to steel as they will not rust.

The surface of the baseboard is ½in chipboard which is screwed directly to the softwood frame. Care should be taken to ensure that the surface fits the frame closely; this is of paramount importance on the short edges as they will need to butt up accurately to adjoining baseboards when the layout is erected in order to carry the tracks across.

The baseboards are joined together by the use of brass hinges screwed into the side members of each baseboard; the head of the hinge pin is filed off and the pin removed so that the baseboards can be separated. A replacement pin is made from a similar diameter brass rod to maintain a close fit, and is bent over to form a handle. This arrangement has been illustrated in earlier chapters.

The provision of a backscene is easily achieved, being simply pieces of ⅛in plywood about 8in high screwed to the backs of the baseboards. It is a matter of personal choice whether this is carried around the two extreme edges of the layout or not. If it is, the inner corner of the joint should be rounded for appearance and effect.

Trackwork

The trackwork for this layout will be

largely sunk in the surface of the yard and the use of Peco flexitrack, for instance, would be ideal. For the more fastidious, there is a finer scale trackwork produced by Shinohara for the American market which can be purchased in the UK from specialist American and continental model railway dealers. It is reasonably priced and an extensive range of pointwork is also available, but because it is designed to take the finer-profile American wheels, not all European-manufactured American models, particularly older ones, will run through it. The Peco trackwork will, however, accept all but the coarsest or finest of wheels and for all practical purposes is ideal.

Tracklaying commences only after its final position has been determined and marked on the baseboard. The track is not laid directly on to the chipboard but on to cork sheet underlay which is glued on to the baseboard in the areas where the track is to be laid.

The cork sheet for this layout needs to span the width of all the parallel tracks, because the layout represents a yard area and not a main line and the aim is to show track which is well covered but not deeply built up on ballast. The majority of the track will eventually be covered with walkways, crossing points for road vehicles and light ash or similar dressings.

The major disadvantage of chipboard becomes apparent when the mounting of point motors beneath the baseboard is considered. If the Peco motors are used, a rectangular piece of baseboard must be removed and this will require drilling holes at each corner of the rectangle and joining them with cuts from a narrow-bladed saw (Figure 44). The hole must be no larger

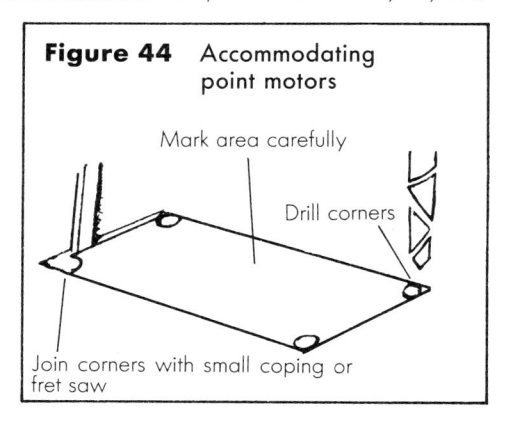

Figure 44 Accommodating point motors

Mark area carefully
Drill corners
Join corners with small coping or fret saw

than is necessary to allow the motor to be clipped beneath the point (or 'switch' in American parlance). There are other point motors available which require only a small hole or slot to accommodate a pin which pivots to move the point blades, but these usually require some form of crank above the baseboard which can be unsightly.

When the cork has dried thoroughly, the track is pinned down commencing with the points and then the plain track in between; ensure that no points lie across the baseboard joints. The track is laid across the joints and then cut with a razor saw. For added strength, small woodscrews are fixed into the baseboard immediately beneath the rails on either side of the baseboard joint to which the rails are soldered (Figure 45). For maximum strength, the screws should be long enough to go through the chipboard and into the softwood frame; $\frac{3}{4}$in screws should be long enough if $\frac{1}{2}$in chipboard has been used. The plastic sleeper at this point is removed but may be replaced for cosmetic purposes once the soldering is complete. This is also the case where plain track is joined either to adjacent plain track or to point work to enable

An American 'shortline'

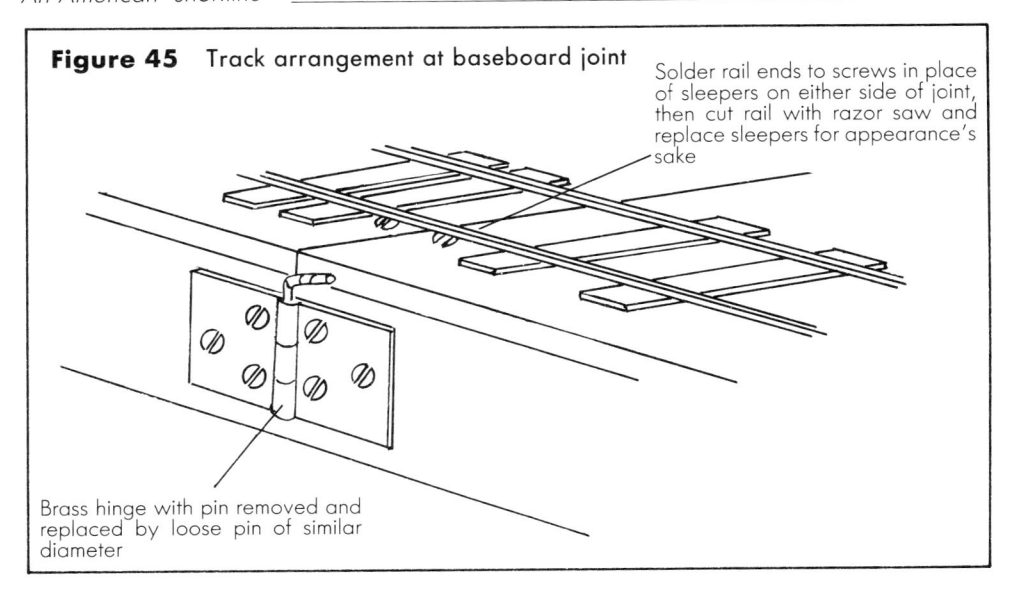

Figure 45 Track arrangement at baseboard joint

Solder rail ends to screws in place of sleepers on either side of joint, then cut rail with razor saw and replace sleepers for appearance's sake

Brass hinge with pin removed and replaced by loose pin of similar diameter

the fishplates or rail joiners to be slid on to the rail. Keep these sleepers (or 'ties' as they are called in America) because once the track is joined and pinned down they can, once the moulded rail chairs have been nicked off with a craft knife, be slid back under the rail and glued or pinned in place to improve the appearance.

The layout should now be tested thoroughly and any adjustments made; rail joints should be smoothed with wet-and-dry paper or a small file. Ballasting is left until most of the scenic details have been added.

Wiring

Wiring the layout is straightforward. Channel the wiring in harness along the underside of the baseboards to multi-pin plugs and sockets leading to a demountable control panel.

If one of the more sophisticated electronic controllers requiring a separate transformer is used it is wise for reasons of safety not to feed mains power directly into the control panel. The transformer should be housed in a separate box, clearly marked 'Danger — Mains Voltage', and the box should contain a number of small holes for ventilation. It can then be placed beneath the layout and connected to the mains and to the layout on a safe low voltage to control the trains and accessories. Needless to say, the specific instructions supplied with the equipment should be followed. If in doubt, a reputable model railway dealer or the manufacturers themselves will advise on suitable equipment, and a qualified electrician will advise on installation and mains connection.

The wiring for this layout (Figure 46) consists of only three switched feeds to the track but contains a number of sections where locomotives may be stored or held while others are moved on the same track. This is achieved by connecting simple on/off switches to a gap in

Figure 46 Wiring diagram

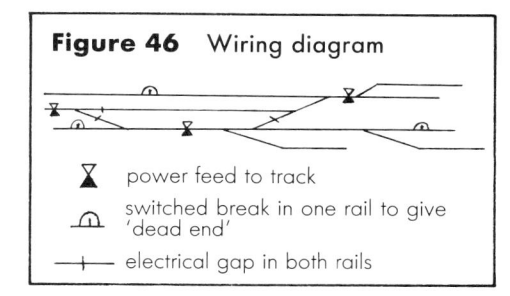

▼ power feed to track
⌒ switched break in one rail to give 'dead end'
—|— electrical gap in both rails

one of the running rails which breaks the electrical feed to a section of track when switched off.

Scenery

The scenery for this layout is simple and principally involves warehouses and industrial facilities at the rear (Figure 47). The backscene, or at least the main part of it, need only represent sky; it can either be covered in sky paper or painted a light blue/grey. If the paper is used, it is essential to apply it carefully and smooth out all air bubbles and wrinkles. Nothing looks worse than backscenes with bubbles and lumps where dust and other debris has been left trapped behind the paper. If the backscene is carried around the two ends, the corner will need to be disguised either by a thick cardboard inner curve or by the addition of polystyrene coving (Figure 48). Whichever is used, the edges should be carefully sanded and smoothed down to blend unobtrusively into the backscene itself. Paper or paint is then applied in the normal way. The ends could be painted or covered with cutouts from other background sheets to represent either a continuation of the industrial scene or open country; if the latter is chosen, a few simple brush strokes will easily convey a flat, midwestern landscape with distant mountains. When the backscene is complete, a coating of clear matt varnish will help to protect it and, if paper has been used, seal it.

The main structures are industrial buildings and warehouses, cattle pens and an oil terminal. The buildings are almost entirely along the back of the layout, running the whole length of the visible portion. One of the advantages of American industrial buildings is that they tend to be flat-roofed and consequently they can be made quite narrow

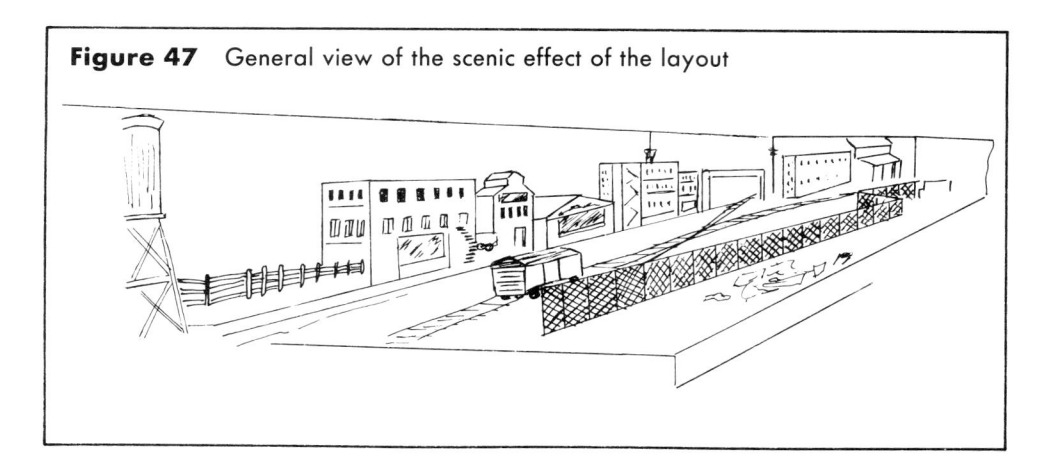

Figure 47 General view of the scenic effect of the layout

An American 'shortline' 79

Figure 48 Painting the backscene

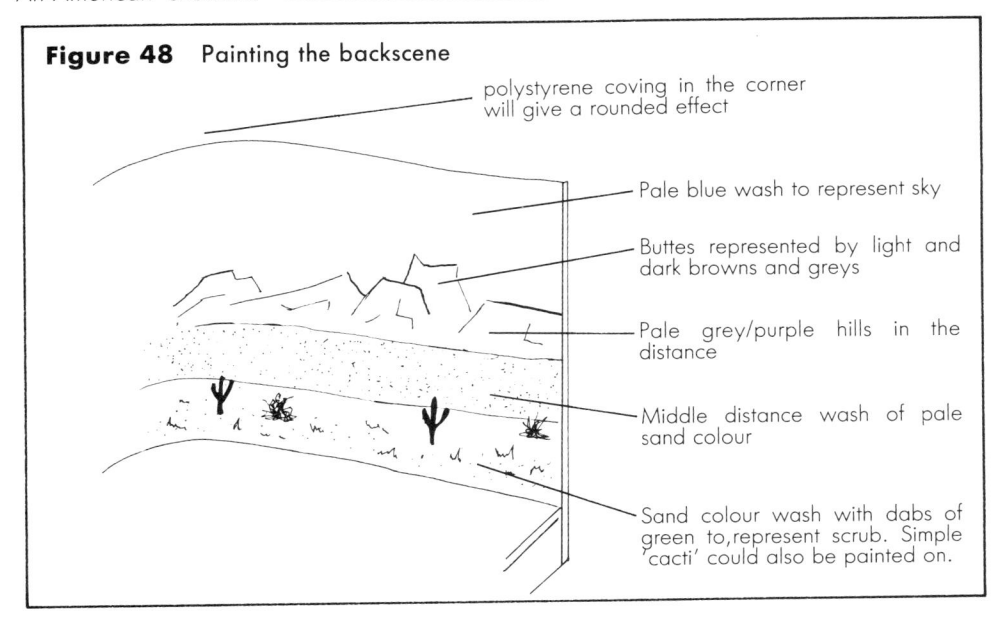

- polystyrene coving in the corner will give a rounded effect
- Pale blue wash to represent sky
- Buttes represented by light and dark browns and greys
- Pale grey/purple hills in the distance
- Middle distance wash of pale sand colour
- Sand colour wash with dabs of green to represent scrub. Simple 'cacti' could also be painted on.

in low relief without creating the perspective problem that arises with similar treatment of a pitched roof. Here you have a choice between making narrow buildings such as this in low relief and perhaps squeezing in an additional siding at the front or modelling the buildings in greater depth, perhaps adding detail around the sides and edges. There is also a choice as to how the main buildings can be provided. In the plan, a cattle pen, a meat packing

Small warehouse-type buildings of more modern architecture make a contrast to ornate Victorian-style multi-storey structures.

works and a series of general warehouses are shown, looking from left to right, and the fiddle yard is screened by an overhead gantry carrying pipework immediately adjacent to which is another large building also screening the fiddle yard which could be a brewery or the back of another warehouse. Finally there is the diesel depot/workshop in front of which is the oil terminal and its pipework and connections leading to the storage tanks, which it is assumed are either 'off stage' or underground.

The front of the layout is scrubland with the usual debris of the modern age scattered around, together with the odd bit of vegetation and a chain link or weld mesh fence on concrete posts.

Starting at the back first, the main *warehouses* can be either scratch built quite cheaply and effectively from simple materials or they can be made from a number of the American outline plastic building kits of warehouses, breweries and other similar buildings. You can, of course, get two for the price of one by literally cutting them in half, as only the frontages are required in this low relief arrangement. Combining bits from different kits, or 'cross-kitting' as it is sometimes called, will give some variety and individuality. It will be necessary to cut loading bays into some of the buildings and to provide crates, barrels, packages and other goods, not forgetting some workmen. For the meat packing works, carcases could be made from Milliput or similar material, or plastic cows could be cut, painted and hung on butchers' rails just inside the loading bay.

As always, the buildings will repay careful painting and weathering. The addition of lettering on the brick can be accomplished by painting panels, in say, yellow and adding rub-on lettering of your choice in black or white. The whole of the brick buildings should be painted a matt colour to represent the main shade of the brick and individual bricks and courses picked out in slightly lighter and darker shades. 'Dry brushing' is a useful technique where a brush is loaded with a particular colour which is then almost completely wiped off and the brush is rubbed across an area to leave light smears of colour. This method could be used around drain-pipes to indicate moss growth from leakages, and black or very dark blues and browns which are almost black could be applied under windows to show staining and discoloration from drips. Cement mortar can be represented by using a very diluted grey paint and flooding it over the brick, then wiping it off the face of the wall with a clean cloth whilst still wet, leaving a residue in the embossed mortar courses. This should, of course, be done before other detail weathering and painting is undertaken. Modern concrete block or concrete clad buildings would make a nice contrast to the brick buildings and kits are available.

Sliding doors on the loading bays can easily be added as shown in Figure 49.

The *gantry* carrying the pipework over the railway to hide the entrance to the fiddle yard is easily constructed from plastic girder section and tube available from model shops (Figure 50). Its clearance over the railway should be just sufficient for the maximum loading gauge, that is tall and wide enough to just clear the tallest and widest locomotives and wagons you will use.

An American 'shortline'

Figure 49 Meat packing plant details

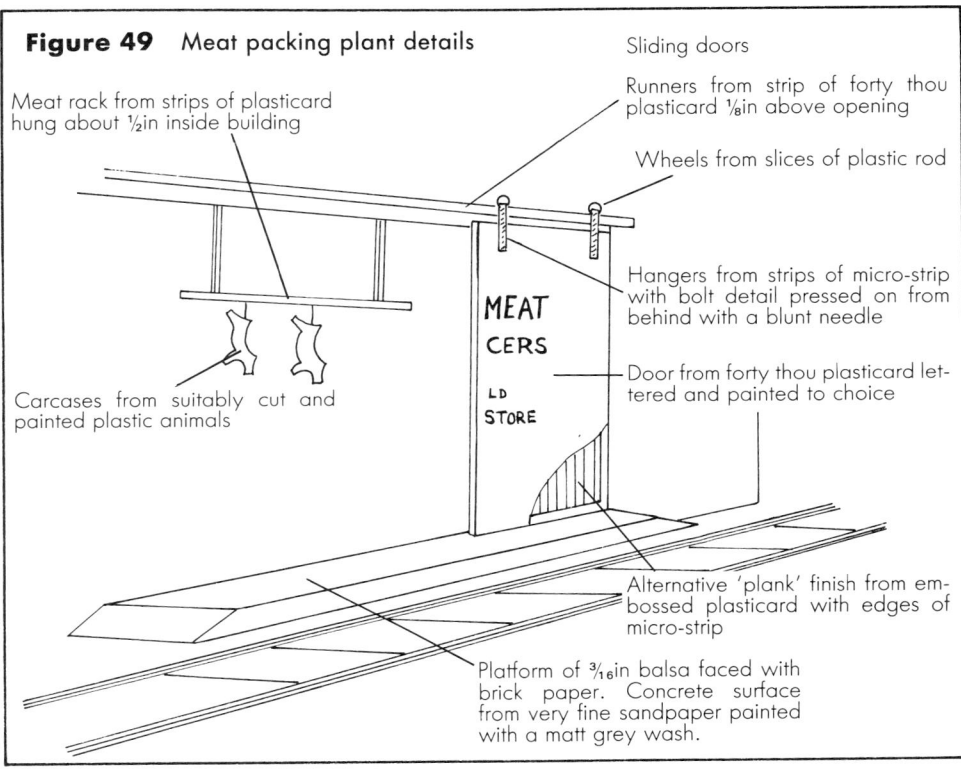

Sliding doors

Runners from strip of forty thou plasticard ⅛in above opening

Meat rack from strips of plasticard hung about ½in inside building

Wheels from slices of plastic rod

Hangers from strips of micro-strip with bolt detail pressed on from behind with a blunt needle

MEAT CERS

LD STORE

Carcases from suitably cut and painted plastic animals

Door from forty thou plasticard lettered and painted to choice

Alternative 'plank' finish from embossed plasticard with edges of micro-strip

Platform of ³⁄₁₆in balsa faced with brick paper. Concrete surface from very fine sandpaper painted with a matt grey wash.

Figure 50 Pipework gantry

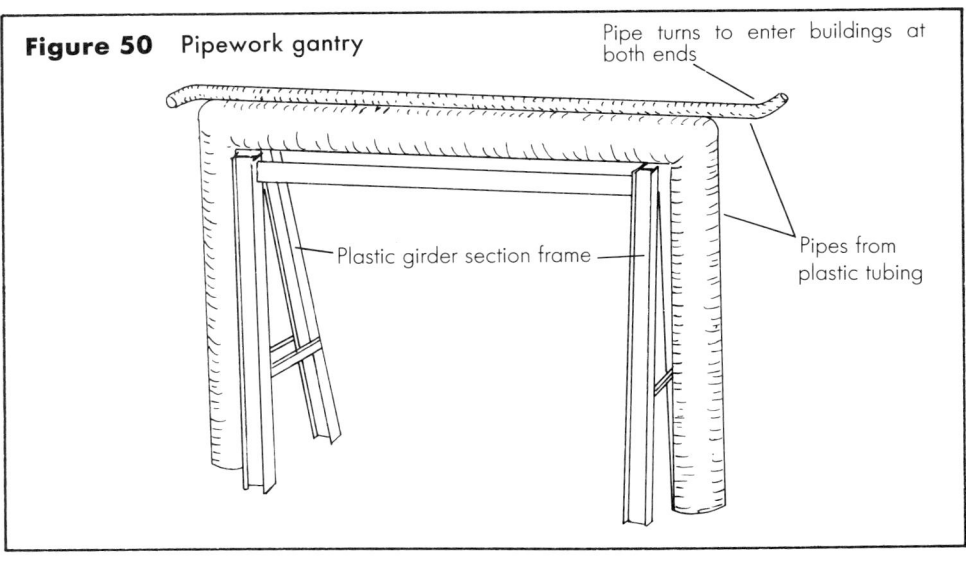

Pipe turns to enter buildings at both ends

Plastic girder section frame

Pipes from plastic tubing

82   Simple model railway layouts

Figure 51   Diesel depot/workshop detailing

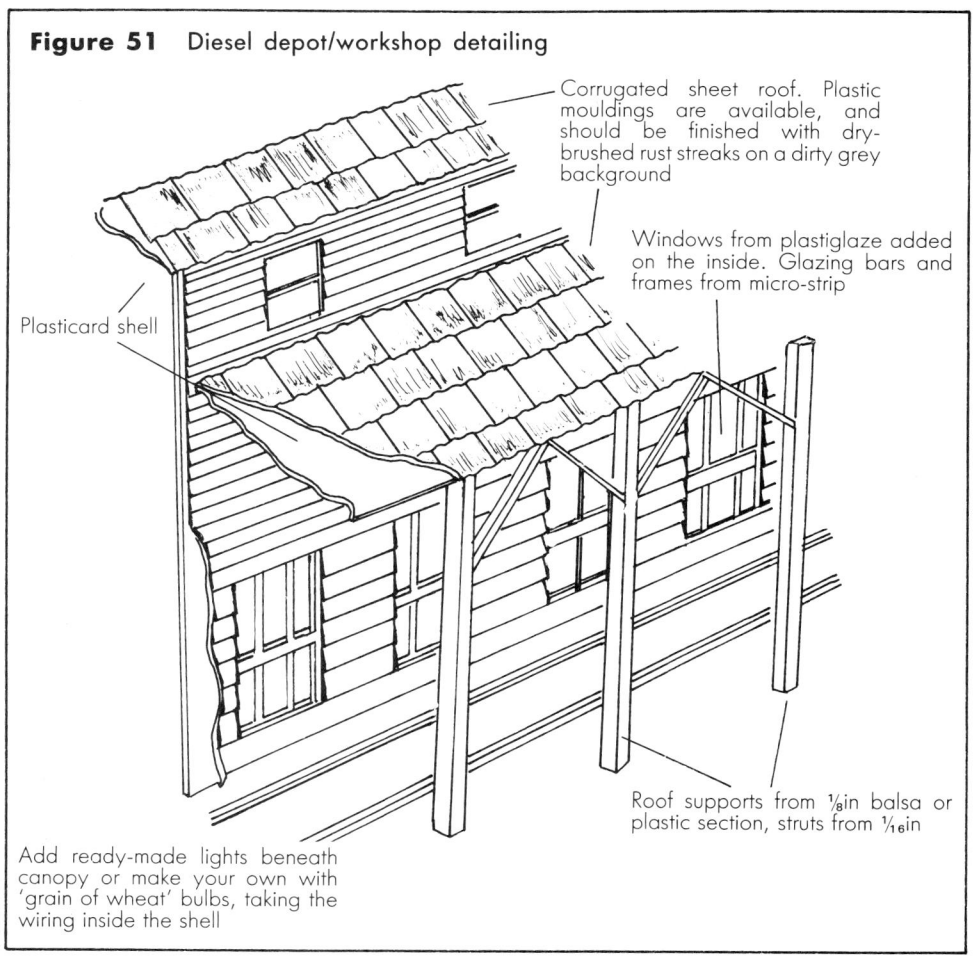

Corrugated sheet roof. Plastic mouldings are available, and should be finished with dry-brushed rust streaks on a dirty grey background

Windows from plastiglaze added on the inside. Glazing bars and frames from micro-strip

Plasticard shell

Roof supports from 1/8in balsa or plastic section, struts from 1/16in

Add ready-made lights beneath canopy or make your own with 'grain of wheat' bulbs, taking the wiring inside the shell

Figure 52   Cattle pens

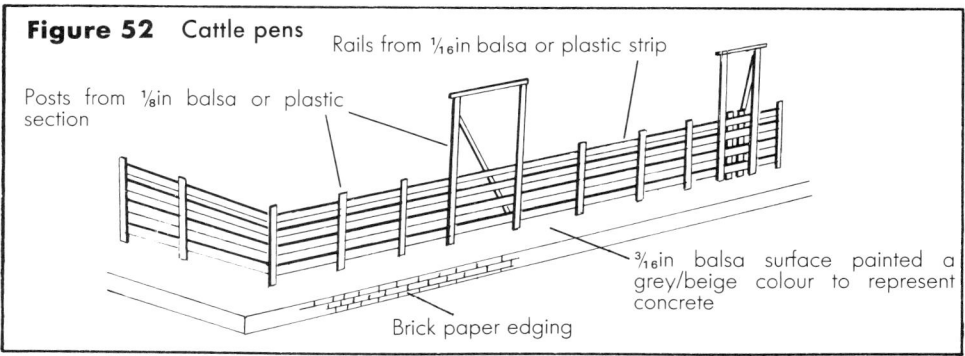

Rails from 1/16in balsa or plastic strip

Posts from 1/8in balsa or plastic section

3/16in balsa surface painted a grey/beige colour to represent concrete

Brick paper edging

An American 'shortline'

The other *warehouse* which hides the fiddle yard is again modelled in low relief and all that is required is the representation of the side and rear of an old industrial building. No direct access or loading bays to the railway are required. A fire escape on the end wall would be a nice detail and could either be borrowed from another kit if plastic kits are used or fabricated from suitable plastic steps or ladders and plasticard strip.

The *diesel depot/workshop*, such as it is (Figure 51), consists of one side, one end and a roof. It would most probably be a ramshackle affair, possibly of wooden construction, which would make a nice contrast to all the brickwork used elsewhere. If interior lights were used, interior detailing could be added, such as an office and a work-bench with a variety of tools, jacks and gantry cranes. There are some excellent packs of details such as oilcans, tools, and other useful paraphernalia available from American manufacturers and a number of specialist dealers in the UK.

The *cattle pens* (Figure 52) are not quite the same as those used in the UK and consist of a series of wooden pens on a low concrete base not much higher than rail level. This structure could easily be constructed from a balsa wood platform with rails and posts of fine balsa or plastic rod if they are of the tubular steel type.

Finally, construction details of the simple *oil terminal* feature are shown in Figure 53.

Ballasting of the track has been left to this stage because it is an effective part

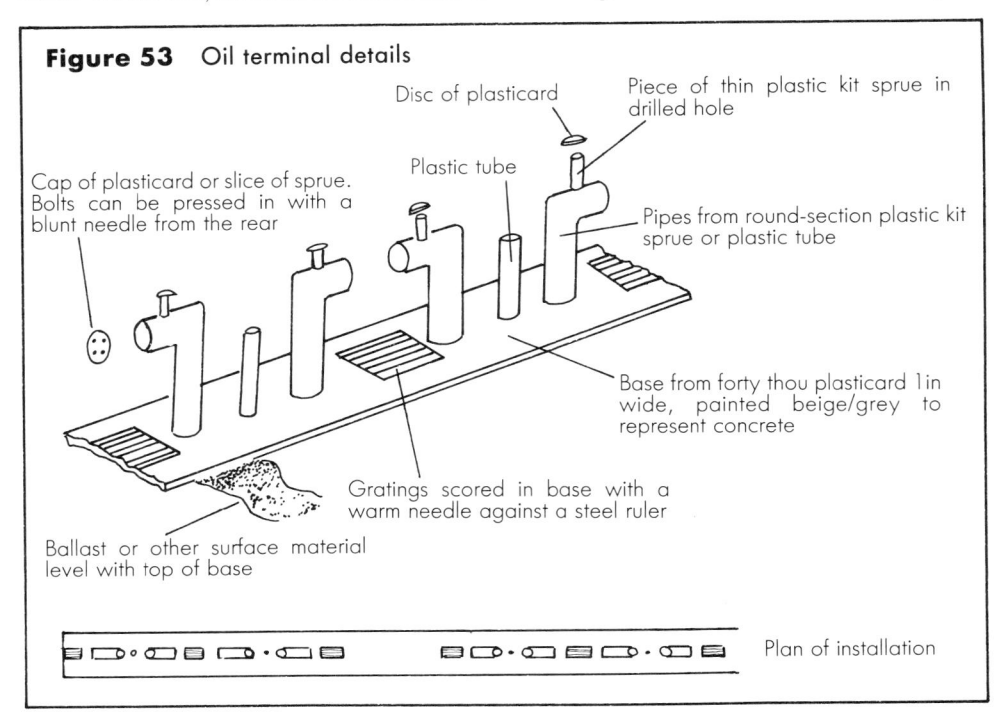

Figure 53 Oil terminal details

- Disc of plasticard
- Piece of thin plastic kit sprue in drilled hole
- Plastic tube
- Cap of plasticard or slice of sprue. Bolts can be pressed in with a blunt needle from the rear
- Pipes from round-section plastic kit sprue or plastic tube
- Base from forty thou plasticard 1in wide, painted beige/grey to represent concrete
- Gratings scored in base with a warm needle against a steel ruler
- Ballast or other surface material level with top of base
- Plan of installation

of the scenery and the treatment of the track will depend to a certain extent on the type and use of the buildings. The whole track needs to be ballasted to a consistent level, and the ballast from Woodland Scenics' range is ideal. It is brushed over the track to the required depth, just below sleeper level; take care that ballast is not left between point blades or in the tie bar area. Once you are happy with the ballast a 50/50 mix of PVA glue and water, with a dash of washing-up liquid added to break the surface tension, is carefully applied over the ballast; to avoid disturbing it, use an eye-dropper or similar. This is a laborious but necessary task. The ballast should be left for 24 hours to dry and then any surplus can be gently brushed away and collected in newspaper for re-use. Any bald spots which occur can be re-treated in the same way and left for a further 24 hours to dry.

Detailing can now begin. The sides of the rails should be painted a rust colour and when this is dry the whole area of the trackwork can be painted with a diluted mix of track colour. This can be airbrushed or applied with a broad $1\frac{1}{2}$in brush.

When this is dry, the area around the cattle pen should be treated to represent straw and fodder which has spilled on to the track. This is not easy to represent but is worth the effort. The most effective method is to use actual dried grass, choosing the finest possible; the 'feathers' from certain types are most effective. The grass should be cut to lengths of 2–4mm and split several times to make it even finer. Then it should be glued liberally around the track near the cattle pens, inside the pens themselves and especially between the rails and the pens. Alternatively, the dried remains of the contents of camomile tea-bags could be used.

The area in front of the meat packing

This picture shows the early stages of making roads around buildings and between trackwork, using card sold for picture mounting.

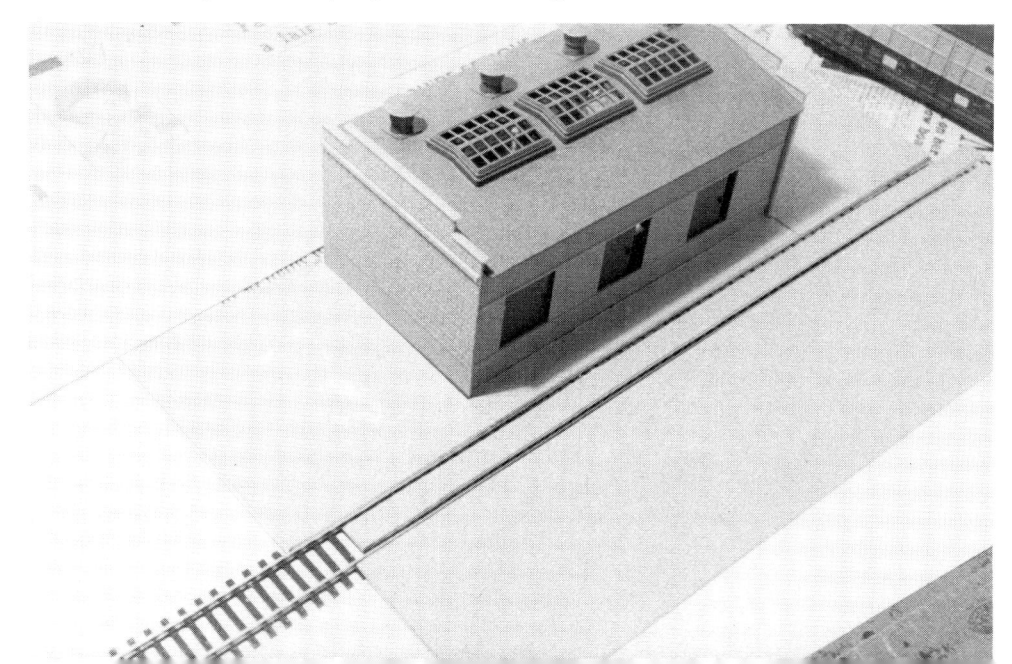

works, and any of the others you choose, may also require access for road vehicles. The level of the road between the buildings needs therefore, to be built up to rail level. A piece of card sloping up from the surface of the baseboard to rail level will do the trick, and it can be covered or painted to represent your chosen road surface. The traditional method is to coat the card liberally with a suitably-coloured matt emulsion and sprinkle talc or other fine powder over it whilst wet. Alternatively, use a large brush to paint on a thin coat of Polyfilla which has been coloured at the mixing stage with watercolour powder paint. The surface can be continued between the rails, allowing a gap between the card and the inner edge of the rail to accommodate the flanges of the rolling-stock (Figure 54). A similar method could be used to continue a roadway towards the front of the layout if desired.

The area around the diesel depot/workshop can provide a great opportunity for those who like to detail models. Firstly, the surface of the track and its surroundings should be brushed with a diluted matt black or very dark grey paint, and when this is dry the area of track a few inches outside the workshop should be liberally coated with a diluted gloss black paint with a touch of blue added to represent the seepage of diesel fuel. Rust-painted barrels, tools, old grindstones and general debris can be liberally scattered around. Many of these details are available from American manufacturers, and the British ranges of Springside, Dart, GEM and Langley, to name but four, yield a useful source of detailing parts from cast metal tools to old boilers.

The front edge of the baseboard is finished relatively easily but can still add a lot of interest to the model. The ground is made up from a thick and uneven coating of Polyfilla, coloured during its mixing with watercolour powder paint to the earth colour of your choice; a sandy colour would probably be best for the western United States. Before this is dry, highlight some areas with different tones, push

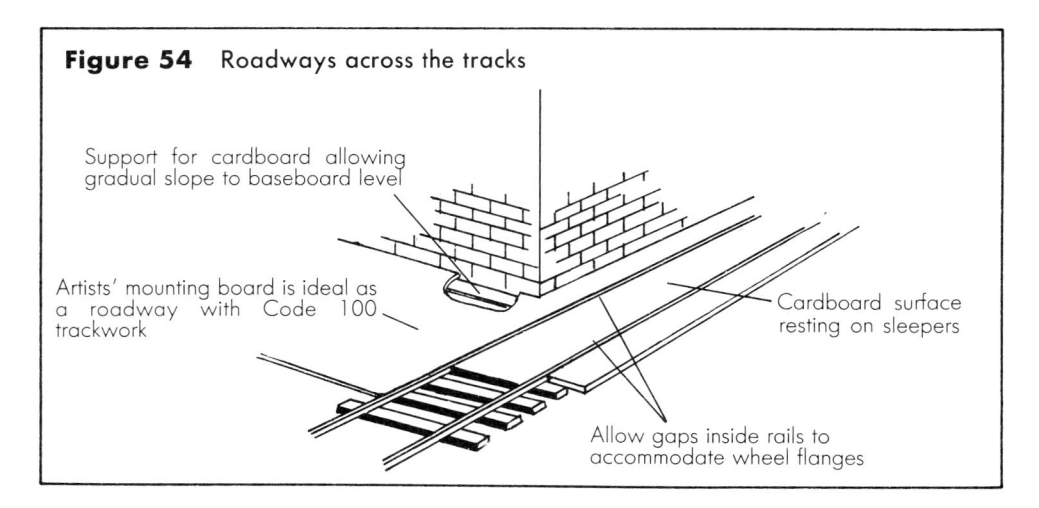

Figure 54 Roadways across the tracks

Support for cardboard allowing gradual slope to baseboard level

Artists' mounting board is ideal as a roadway with Code 100 trackwork

Cardboard surface resting on sleepers

Allow gaps inside rails to accommodate wheel flanges

The 'burnt-out' car and wire netting fence, as described in the text. The car is an old child's toy, picked up for a few pence at a jumble sale. The fencing is made from fine netting of the type sold to dressmakers for making veils on ladies' hats, and is supported on balsa wood posts. Alternatively, more expensive etched fencing of various types is available from John Piper for those who want a 'Rolls-Royce' finish.

one or two small stones and boulders into the wet mix and plant a few bushes made from teased-out Woodland Scenics foliage material. When dry, drill some holes 1–2in from the front edge of the baseboard about 3–4in apart to accommodate balsa wood fenceposts $\frac{1}{8}$in square, 4in long with $3\frac{1}{2}$in protruding from the baseboard. Glue these in and fasten to them the chain or mesh link fence etching from the Scale Link range. Paint the fencing grey and the posts beige to represent concrete. Bend and twist the mesh in places to show misuse and streak it with rust paint, dry brushing as described earlier.

If you have young children or occasionally visit local church fêtes, bazaars or rummage sales, you should be able to acquire a battered die-cast American car of roughly HO scale for a few pence. This can be further vandalised, a good proportion of its remaining bodywork painted a rusty red and dumped on the wasteland!

Some semblance of vegetation can be added, made from little clumps of turf and grass mixes, the lighter greens being most appropriate, and a tree could be added. Woodland Scenics produce some nice tree kits and even some little 'cameo' scenes surrounding the trees. Final detail depends on the individual but should include some road vehicles. There are plenty of cheap children's toys of American-type cars and lorries at approximately HO scale to populate this layout. A little judicious repainting and weathering improves their appearance no end, and the people themselves can be added from any of the ranges of HO scale figures.

Locomotives and rolling-stock

The range of American outline locomotives and rolling-stock available in the UK is quite considerable. So far as the locomotives are concerned, the smaller, older types of diesel are ideal. The colours and livery are not important as they can be repainted or assum-

An American 'shortline'   87

Above *The layout suggested could easily be built to represent the steam age and small 'switchers' such as the one shown here would be ideal motive power. The weathering effect on the locomotive is worth noting, and is far more common in America than in Britain. Superb results can be obtained and articles on the appropriate techniques appear from time to time in the American model press.*

Below *The Americans produce an extensive range of kits for locomotives, rolling-stock and structures, varying from the 'craftsman' type to relatively straightforward and simple ones. MDC Roundhouse kits are well known and readily available in Britain. The caboose kit shown here has a pre-painted body with detail fittings which are added by the builder. Complete locomotive kits are also available at a reasonable price in both HO and HO N3 narrow gauge.*

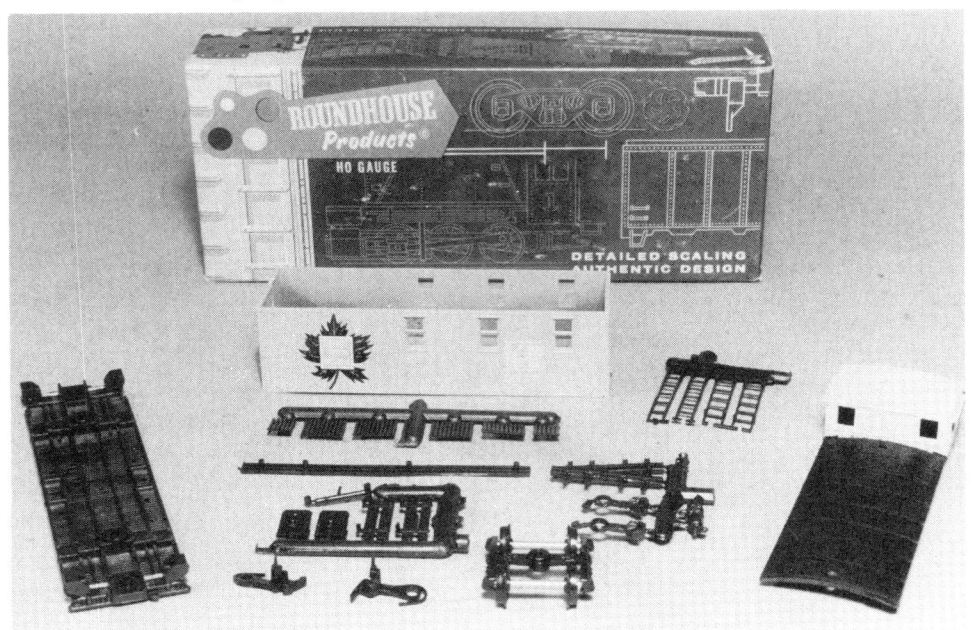

Simple model railway layouts

Left *Two contrasting boxcars made from the ready-painted Roundhouse kits.*

Below left *Large by British standards, but on the small side by North American standards, this Southern Pacific 2-8-2 is a beautifully-detailed but quite cheap model typical of many available for the American market; this is a Bachmann model. If you can afford them, hand-crafted brass locomotives made by Far Eastern manufacturers are available and offer museum quality detailing.*

ed to be leased, loaned or recently purchased from another main line company. Detailing kits, decals (transfers) and a host of conversion parts for ready-to-run models are available from specialist retailers in Britain. Regular articles appear in American magazines which can be purchased here as well as British magazines such as *Continental Modeller* and *Scale Model Trains*; these often contain articles on American models, and detailing and converting locomotives and stock.

The choice of industry on the layout will determine the types of vehicles required. The layout will certainly require stock-cars, reefers (refrigerated box-cars) for the meat packing plant, oil-tanks for the oil terminal and general box-cars for other uses.

Further references

For those interested, there are quite a lot of books and magazines available in the UK on both model and prototype American railways as well as occasional articles in the British model railway press. Some are listed below.

Magazines

Model Railroader, a US magazine available from specialist importers and on subscription.

Narrow Gauge and Shortline Gazette, similarly available.

Continental Modeller and *Scale Model Trains* both have occasional articles on US modelling and railways.

Books

Amtrack, R. Bradley, Blandford Press

The Model Railroading Handbook (several volumes), Schleicher, Chilton Book Co

CHAPTER 6
009 narrow gauge quarry layout

This layout goes some way towards satisfying the needs of those who like a lot of railway and activity in a small space. To do so it makes use of two factors. Firstly, there is the ability of narrow gauge railways to run and operate effectively round sharp curves and to climb fairly steep hills with by model standards, a respectable load. Secondly, there is the intensity with which industrial railways, often narrow gauge, occupy fairly small sites. These two characteristics can be used to advantage to provide the basis for the layout illustrated in Figure 55.

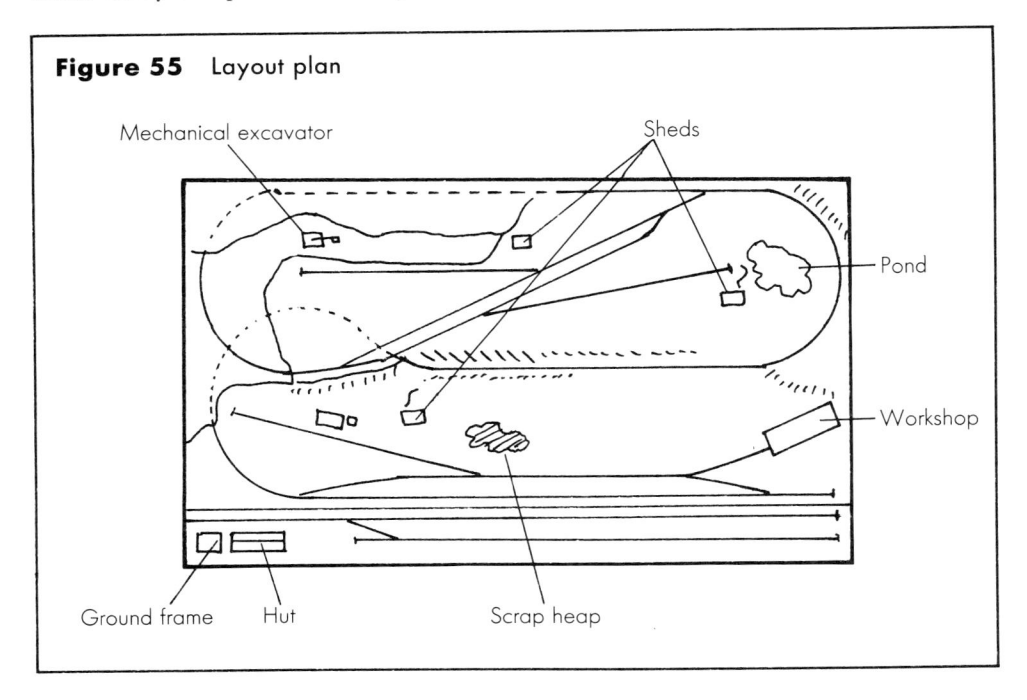

Figure 55 Layout plan

The plan
The layout is based on a mineral working, in this case a chalk pit, and a loading bank or quay whereby the mineral extracted is transferred to standard gauge railway wagons for distribution. It does not, however, attempt to provide an accurate representation of a real chalk pit.

The basic operating scenario for the narrow gauge system would be the bringing of filled wagons from the quarry face to be loaded into the standard gauge wagons, and the return of empty wagons. It is, alas, not possible to arrange for the chalk to be loaded and unloaded realistically, so I am afraid that your imagination must be relied upon. It would be possible to run two trains on the layout by holding one train in the siding and one in the loop, but I suspect that one would be sufficient.

The standard gauge sidings give the opportunity for further development in that they allow for the model to be incorporated later into a standard gauge layout. Alternatively, this layout could be extended; the standard gauge sidings could be simply connected to a fold-down or added-on fiddle yard to enable standard gauge trains to be run in and out of the transfer sidings or an additional board could be added which could provide scope to model perhaps part of a village or an extensive farm complex together with some more standard gauge sidings to allow shunting to take place on the two boards.

A small industrial layout of this type offers a lot of modelling and operating potential in a very small space. It also offers considerable scope for those interested in scenic modelling and for those with a mechanical bent who could take great delight in automating the running and mechanizing of the mechanical shovels, excavators and other paraphernalia which are involved in such industrial activity.

The layout could also be readily adapted to suit your own ideas. For example, the standard gauge track could be replaced by a canal, and the chalk by a sand quarry. Above all, it provides lots of chances to experiment scenically and to develop techniques and ideas.

Baseboard construction
The baseboard can be made in one piece; the plan shown fits into an area 3ft 6in by 2ft. It has a 2in × 1in softwood frame and the construction methods are illustrated in Figure 56.

The softwood frame should be square and flat, and on to it is fixed plywood or chipboard decking at three levels with ramps to accommodate the various tracks required to take the railway from the level of the exchange sidings at the bottom to the quarry face at the top. As with all multi-level layouts, construction should start at the bottom, in this case at the front with the standard gauge track bed. This is made from a piece of chipboard or plywood 4½in wide which is simply screwed flat down on the softwood frame, flush to its front edge.

Immediately behind this is a length of 1½in × ¼in softwood, edge down, which is screwed upright onto the back edge of the standard gauge track bed. A similar strip is also pinned some 9in behind and parallel to this to form the rear support for the second layer of decking. The board which forms this, the lower level of the narrow gauge, is now screwed to these strips; ¼in chipboard or plywood would be ideal.

009 narrow gauge quarry layout ──────────────────────────── *91*

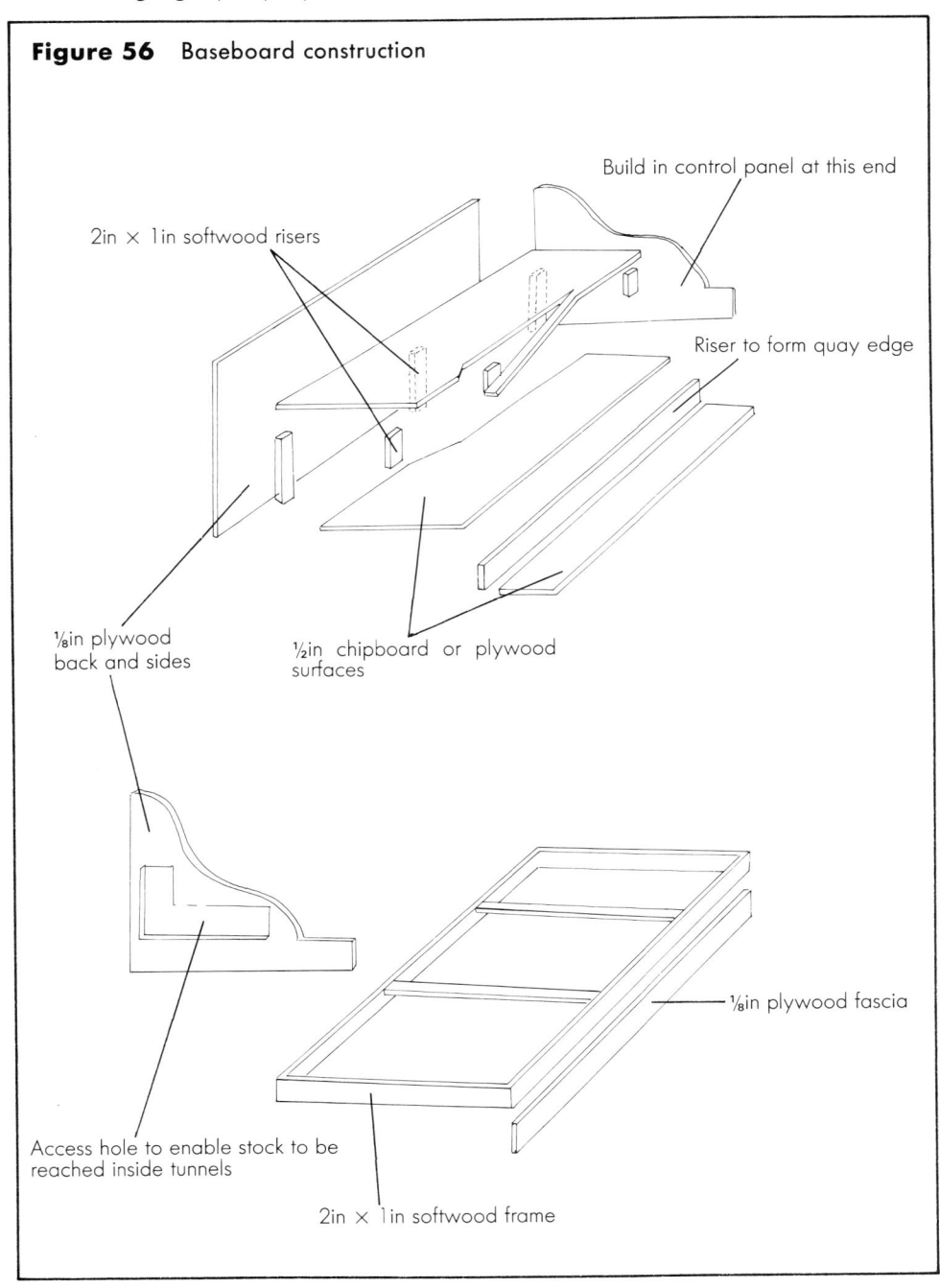

Figure 56 Baseboard construction

Build in control panel at this end

2in × 1in softwood risers

Riser to form quay edge

⅛in plywood back and sides

½in chipboard or plywood surfaces

⅛in plywood fascia

Access hole to enable stock to be reached inside tunnels

2in × 1in softwood frame

The baseboard under construction, clearly showing how the various levels are supported on a softwood frame.

The highest of the three levels is supported on 'stilts' or 'risers' of softwood which are screwed to the softwood frame. These need to lift the decking at least 2¼in to allow adequate clearance for locomotives and stock passing underneath. This board needs to overlap the lower level so that a recess can be cut out to accommodate the gradient which will link the two levels, as shown in Figure 56.

The final stage of the baseboard construction is the addition of ⅛in plywood fascia to the back and sides of the layout to provide a base on which the scenery can be built. The right-hand end also provides the housing for the integral control panel (Figure 57) and it would also be wise to cut an access hole in the left-hand end to enable

This view shows the link between the two levels, and how the decking of the top level has been cut and bent down to provide the incline. The hardboard surrounds will later be cut to the profile of the finished scenery. The point motors are mounted above the baseboard but will be hidden by removable scenic sections and buildings.

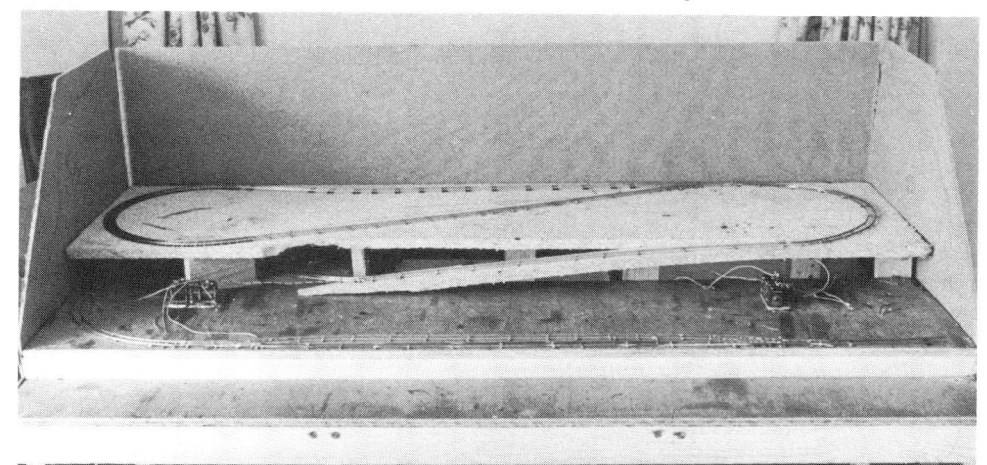

any stock which derails or stalls in the tunnels to be rescued. The plywood fascias could be painted to give a more presentable appearance to the layout, especially if it is to be used in the living area of the home!

Trackwork

The trackwork for the layout can be divided into two distinct parts, the narrow gauge (009) and the standard gauge. Track for the narrow gauge can be made easily and cheaply using rail soldered to copper clad sleepers in the time-honoured way. N gauge rail with sleepers placed randomly would be ideal. However, most builders will probably want to use ready-made flexitrack and Peco produce one in 009 with matching pointwork which has random sleepering and looks effective. The normal range of Peco trackwork accessories such as point motors, buffer stops, dummy point levers and even a wagon turntable can be used to complement the track and are readily available.

Because the curves have a sharp 6in radius, great care is needed to ensure that the track is laid carefully if good running is to be achieved. The curvature must be smooth and even and it is best to avoid rail joints in the curved sections if at all possible. A template can easily be made from scrap ply or hardboard, as shown in Figure 58.

The Peco 009 track is laid in the same manner as other flexitrack, and is pinned down directly to the baseboard surface. Care is needed to ensure smooth joints at points, with no sudden changes of direction in the track leading into and out of them. Ensure that the fishplates are a tight fit and that there are no burrs on the rails at the

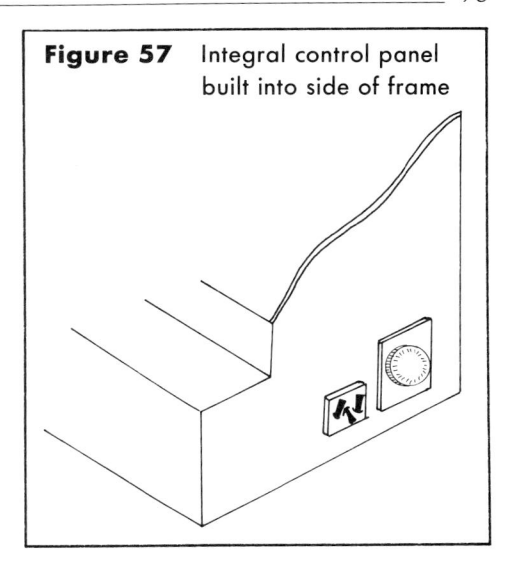

Figure 57 Integral control panel built into side of frame

joints. Aim for accurate alignment and smooth changes of direction; time and care spent at this stage will be rewarded later with better running.

Although there is little standard gauge track, it is all concentrated at the front edge of the baseboard so it is worthwhile making it as fine and accurate in appearance as possible. This is definitely a case for the 'fine scale' bullhead rail trackwork. As there is only one point on the standard gauge line, it would be worthwhile building one from the scale components available. One of the 'fine scale' track systems would be ideal. Ready made-crossing vees and machined point-blades are available to make construction even easier. Taking your time and following the manufacturer's instructions, it is quite a straightforward matter to produce superb pointwork, and there is no easier start than the single plain turnout required here! However, it would not be too expensive to purchase one from the specialists who advertise in the

Simple model railway layouts

Figure 58 Making and using a template for curved track

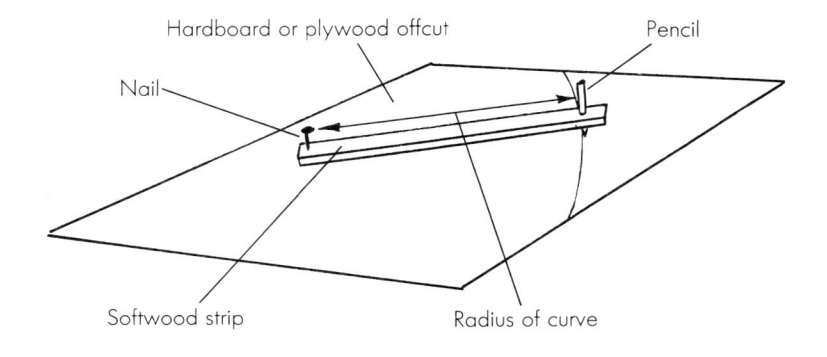

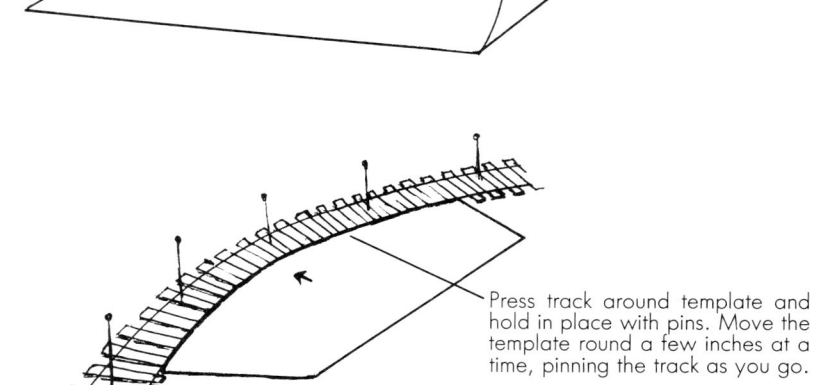

Hardboard or plywood offcut

Pencil

Nail

Softwood strip

Radius of curve

Cut out along marked line and smooth edge with glass-paper

Press track around template and hold in place with pins. Move the template round a few inches at a time, pinning the track as you go.

model railway magazines if you don't fancy building one yourself.

The standard and narrow gauge tracks are laid directly on the baseboard surface. Starting with the standard gauge trackwork, assuming one of the finer flexitrack systems is used, it should be carefully cut and prepared for the layout. The area to be occupied by the trackwork is painted with a 50/50 mix of PVA woodwork glue and water, avoiding the area where the point tie bar will be situated (mark this area when preparing the trackwork). Lay the track and pour ballast over it whilst the glue is still wet. The standard gauge track is likely to be quite decrepit and to represent the spilled chalk dust and, particularly towards the buffer stops, the weeds, apply dry polyfilla powder and 'grass' flock powder with the ballast. When this is dry, the surplus can be collected for re-use.

The narrow gauge track is again laid directly on the baseboard surface. Ballasting is carried out once the track has been laid and tested, and the intention should be to create lightly-ballasted but decrepit-looking trackwork so typical of the industrial narrow gauge railway scene. Use the finest ballast you can find, and brush it into position to your requirements. Again, green scenic dressing can be added to represent weeds and polyfilla powder to represent the chalk. Avoid getting any of this mix in the tie bar area of the points, or in the crossing vees or point blades. When you are satisfied, apply the water and PVA glue mixture (with a dash of washing-up liquid added) sparingly from an eye-dropper. Allow to dry thoroughly and then brush away any surplus or loose ballast. Re-treat any bald patches in the same way.

Wiring

The wiring for this layout is complicated by the addition of a reverse loop. However, it is not too difficult to overcome these problems, and Figure 59 shows the arrangement for wiring

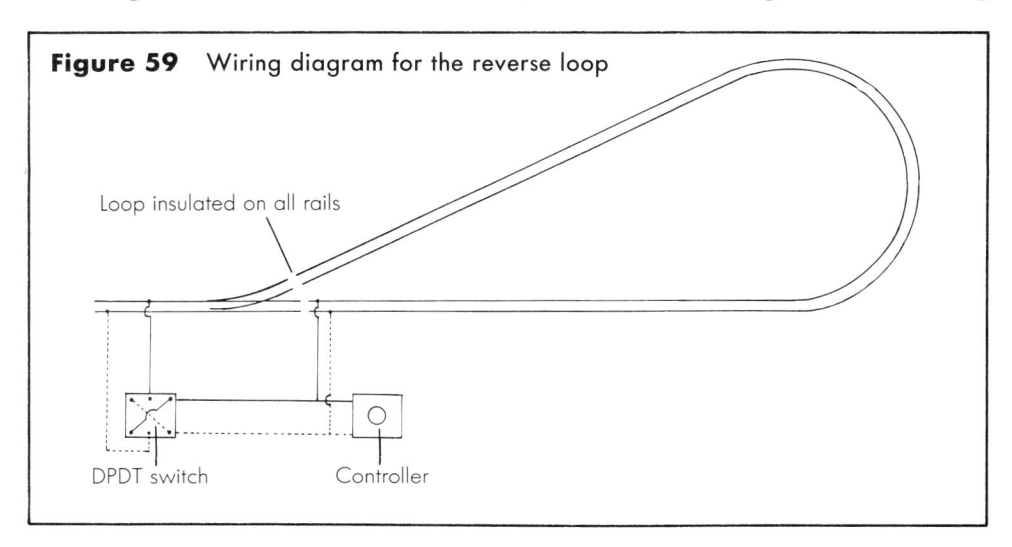

Figure 59 Wiring diagram for the reverse loop

Loop insulated on all rails

DPDT switch   Controller

Figure 60 Wiring diagram

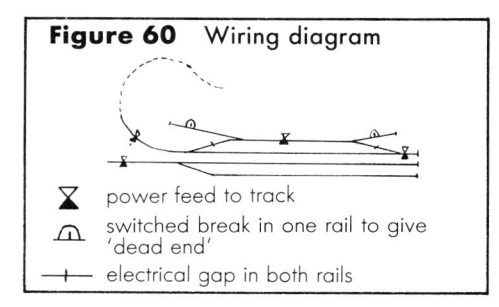

☒ power feed to track
⌒ switched break in one rail to give 'dead end'
—+— electrical gap in both rails

the loop using a DPDT (double pole double throw) switch available from model shops or electrical component retailers. When a locomotive enters the loop, the switch is thrown and the polarity of the current reversed. When the points are changed, the train can leave the loop (the reversal of current does not affect the direction of the

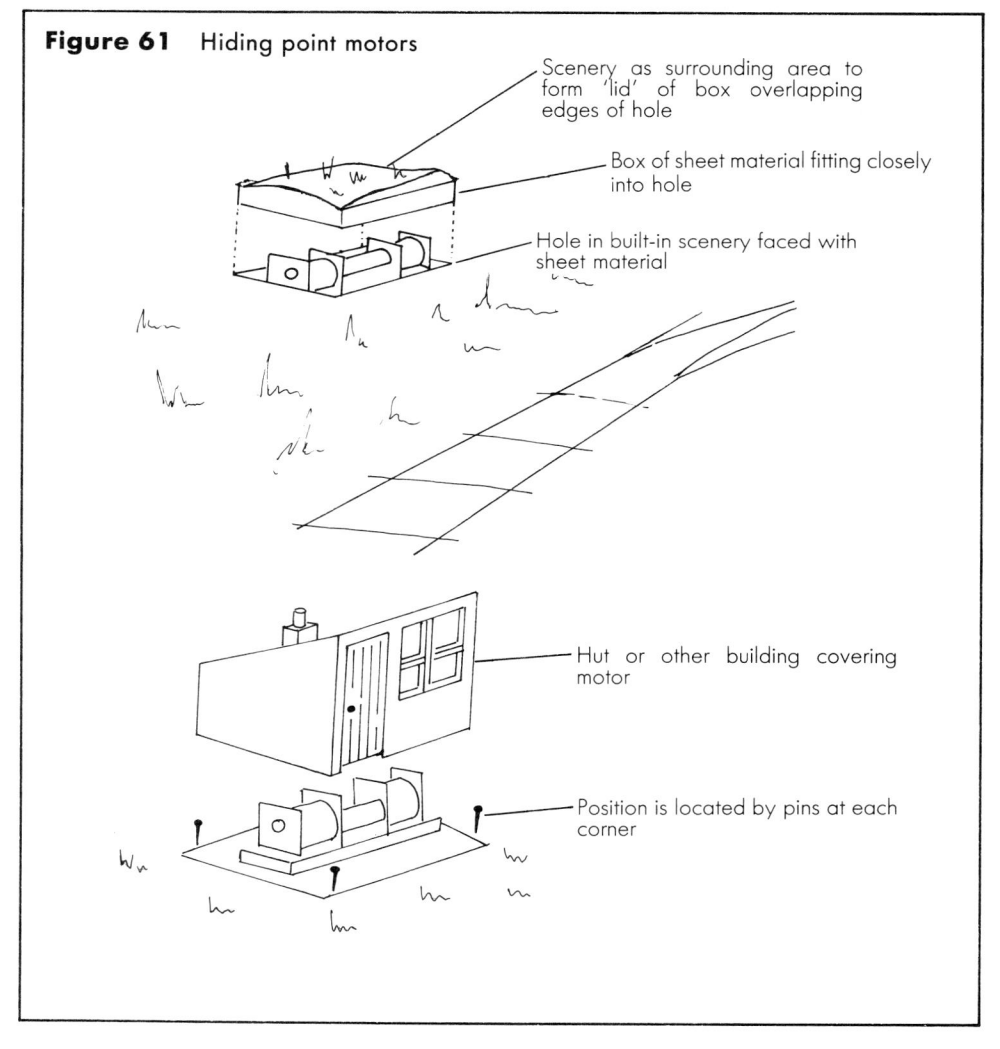

Figure 61 Hiding point motors

Scenery as surrounding area to form 'lid' of box overlapping edges of hole

Box of sheet material fitting closely into hole

Hole in built-in scenery faced with sheet material

Hut or other building covering motor

Position is located by pins at each corner

train because the loop is fed directly).

The layout can be conveniently divided into two electrical sections, the narrow and standard gauge tracks. The standard gauge layout is simplicity itself to wire, requiring just one feed as shown on the wiring diagram (Figure 60). The narrow gauge track requires two feeds, one in the run round loop and another at the head of the loop. The wiring from the trackwork is fed beneath the baseboard to the switches and control panel which are mounted in the side of the baseboard.

Because plywood or chipboard have been used for the baseboard surface, cutting out holes to mount point motors below the baseboard is more awkward than if a softer surface material had been used. If preferred, point motors could be mounted above the baseboard and easily disguised; in a quarry there are usually a great many huts, buildings and rock outcrops under which they could be hidden. A couple of ideas and construction methods are illustrated in Figure 61. The motors should, of course, be installed in accordance with the manufacturers' instructions and the wiring is fed beneath the baseboard to the control panel, where either a passing contact switch or the stud and probe method could be used; if anything, the studs are slightly more compact. Either can be mounted on a miniature track diagram.

Scenery

The scenic work on this layout basically consists of filling in the gaps between the various deck levels to look like rock strata and providing some height at the 'face' on the uppermost level.

Starting at the bottom, the loading platform needs to be formed along the edge of the timber supporting the lower narrow gauge level. The face could be covered in embossed stone or brick plasticard, brick paper or plastic sheet along its length; the stone cappings along the top edge would be quite substantial and could be easily formed from Das or one of the similar modelling clays as shown in Figure 62. The edges of the stones where they join the brickwork below and the surface above should be straight; to achieve this, cut with a craft knife along a straight edge before the compound sets.

Moving up a level, the next task is to fill in the gaps between the lower and upper narrow gauge levels. The easiest method of filling in the narrow gap between the inner edge of the incline and the upper level, without encroaching too much on to the narrow incline, is to drape plaster bandage from the top deck to the inside of the ramp itself. This can later be detailed to represent a rock face using a thin coating of Polyfilla scored to represent the rock strata before it dries completely.

Similarly the area around the lowest tunnel is formed from plaster bandage but this time laid on a framework sculpted from small gauge wire netting stapled to the wooden ends and decks. Aim to have a fairly level top and a sharp, almost vertical, drop to the lower level. Turn in the edges where the track enters the rock face to give an impression of the continuation of rock into the tunnel; again, this can be detailed later when dry. At this stage the idea is to get the basic contours established. Ensure ample clearance for trains entering the tunnel!

98 _____ *Simple model railway layouts*

Figure 62   Modelling the edge of the loading platform

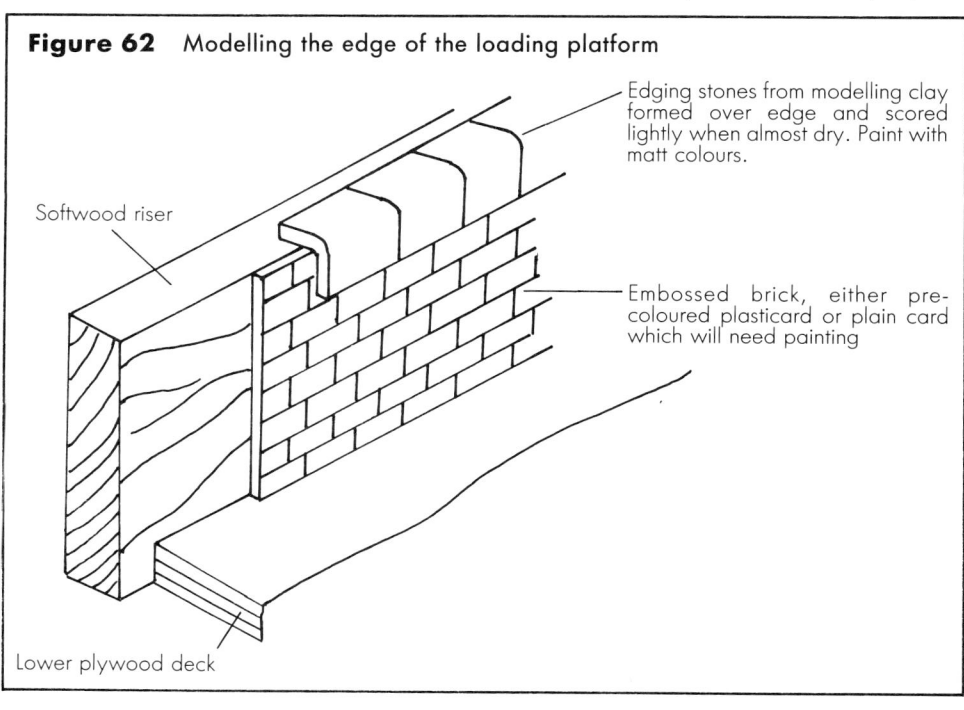

Edging stones from modelling clay formed over edge and scored lightly when almost dry. Paint with matt colours.

Softwood riser

Embossed brick, either pre-coloured plasticard or plain card which will need painting

Lower plywood deck

Figure 63   General appearance of the finished layout

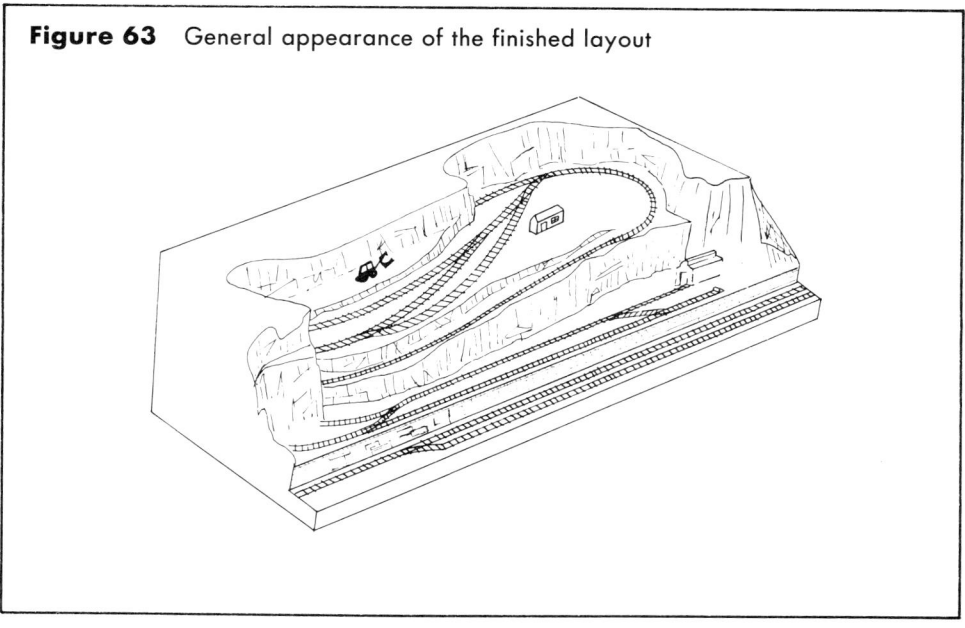

The top deck is the next for treatment and here it is a matter of creating a rock face to hide most of the reversing loop and building up some relief to create an illusion of the landscape continuing beyond the baseboard. Figure 63 shows the general effect which should be aimed for. The main face is built up from layers of polystyrene ceiling tiles laid in such a way as to provide a ledge from which the chalk is excavated and on which the mechanical shovel operates; the tunnels are formed by the simple expedient of continuing the ends of the upper layers over the trackwork. The corners have similarly constructed rock outcrops and the edge boards are given a coat of stiff Polyfilla which, when nearly dry, is scored and sculpted to match the rest of the rock faces.

The bases which have been laid are now coated with a mixture of Polyfilla and a touch of yellow ochre powder colour, and the sculpting of the rock faces is done as the mix stiffens but before it dries completely. The top surfaces should be fairly flat and occasional larger boulders (made from small stones) can be pushed into the wet mix.

Methods of hiding the point motors have already been mentioned, but the larger, flat areas will require a coating of the Polyfilla mix. Some odd lengths of rail and sleeper can be pressed into this mix to represent old trackwork, as can all sorts of debris such as pieces of broken mechanical equipment. Old die-cast toys provide a cheap source of this kind of detailing, and there are plenty of detailing kits which provide buckets, old boilers, sheds, huts, tools and even cats and dogs which can be added later.

The area near the standard gauge line, where bits of open baseboard surface appear, should also receive a coat of thick Polyfilla mix. When dry, this area should be given a coat of PVA glue to which scenic dressings of various colours and textures can be added until you have the desired finish. The standard gauge point motor could be hidden by a hut or even a ground frame. Bushes can be added from foliage mat and some debris once more scattered around.

The flat areas beside the track on the narrow gauge levels are treated similarly and a workshop could be adapted from an N gauge engine shed kit; one of the small wooden continental types made suitably decrepit would be ideal. Tools, oil barrels and perhaps an oil-tank on stilts as shown in Figure 64 could be added.

The flat area at the very top of the layout should be made to represent grass with a coating of a suitable turf mix on PVA glue.

The standard gauge track should have two buffer stops added at the end of each siding and these should be given a thorough coating of rust paint and track colour. This is a tedious job but well worth the effort. The area immediately around them should be covered in coarse turf and foliage mat to represent weed growth. By contrast, the ends of the sidings on the narrow gauge can simply have a sleeper glued across them secured by a chain or rope perhaps made from cotton. Again, treat with undergrowth and weeds.

You can go on almost ad infinitum with refinements and new scenic detail, and, as mentioned earlier, there is no shortage of excellent detailing accessories and parts available.

Simple model railway layouts

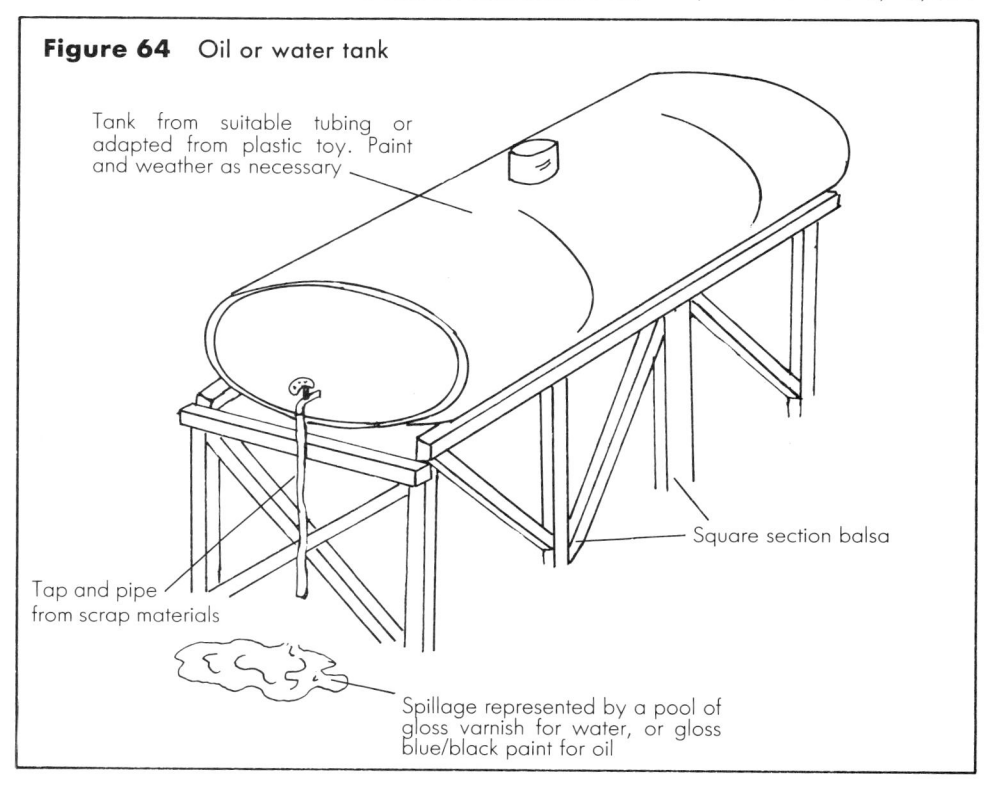

Figure 64 Oil or water tank

Tank from suitable tubing or adapted from plastic toy. Paint and weather as necessary

Square section balsa

Tap and pipe from scrap materials

Spillage represented by a pool of gloss varnish for water, or gloss blue/black paint for oil

The main sources of ready-to-run 009 gauge equipment suitable for an industrial line are Roco Minitrains and the Eggerbahn range which was recently reintroduced by Jouef. The little diesel shown here would be ideal for the quarry layout and is shown with an appropriate short-wheelbase tipper wagon.

009 narrow gauge quarry layout   *101*

The Eggerbahn range includes a number of useful vehicles including the one shown here which would be ideal for the quarrymen's transport.

Locomotives and rolling-stock

The rolling-stock for the layout is easy to select. On the standard gauge, use either a diesel or small steam shunting locomotive depending on the period being modelled, together with a few mineral wagons, either the five-plank or seven-plank wooden-bodied type or the later steel-bodied variety, again depending on period. It would be possible to shunt the standard gauge sidings but this is not intended to be the main operation. It should also be possible to fully automate it, but that is beyond the scope of this book.

Side-tipping wagons and small diesel locomotives (or steam if you prefer) would be ideal for the narrow gauge. Ready-to-run locomotives are available from Eggerbahn and Roco as well as a good many kits from smaller manufacturers who advertise in the model press.

Further references

Narrow Gauge Charm of Yesteryear, I. Peters, OPC

Peco Book of Narrow Gauge, D. Lloyd, Peco

CHAPTER 7
French branch line terminus

In the last few years there has been a tremendous upsurge of interest in foreign railway systems. No doubt this is due in part to a relaxation of the traditional British insularity with the broadening of horizons by TV and increased foreign travel. It may also have just a little to do with the excellent foreign models which are now appearing on the shelves of model shops. Whatever the reason, it is clear that sufficient railway modellers now pursue overseas' railways to justify at least one specialist magazine and regular articles in others. It is also true that the majority of foreign models of whatever gauge are of a very high quality with extremely good detail and usually excellent running properties.

There has unfortunately been a tendency for model railway layouts using this equipment to be extensive and to incorporate the gimmickry of flashing lights, high-speed turntables and all the other toy-like accessories offered in catalogues, which, all in all, are very unrailway-like. Very rarely is all this excellent equipment used realistically and very rarely does one see a small-space foreign model built to the standards of similar British layouts.

So, given the quality of the equipment available and its undeniable appeal, the task is to use it effectively to produce a minimum-space layout, finished to a high and realistic level. Of all the other projects described in this book, this is probably the easiest in which to achieve the desired result, because the basic equipment is so good to start with.

The plan

This project is for a representation of the European scene, and the task is to produce a realistic late steam era layout in a space of 8ft × 1ft.

By far the greatest number of European models produced are of German prototypes, and it would be easy to produce a German layout straight from the box using this equipment. However, models based on other countries' railway systems can easily be produced and this one will depict a French railway.

The two World Wars which have ravaged Europe nevertheless serve the modeller of the post–1945 European scene well. Firstly, they helped to keep in service many antiquated locomotives and items of rolling-stock

This simple but effective scene depicting a branch line train of the pre-war Paris, Lyon and Marseilles Railway in a rural setting, uses readily available materials such as the Jouef building, commercial plastic kit bridge, adapted Rivarossi Bourbonnais locomotive, Piko van and Liliput coach. Note also the careful ballasting of the Peco trackwork which looks completely convincing. (Brian Monaghan, by courtesy of Ian Allan/Model Railway Constructor)

which would otherwise have been broken up many years before. Secondly, and more importantly bearing in mind the preponderance of German steam models, the Germans were forced to provide locomotives and stock to other European railway systems by way of reparation, particularly after the First World War. After the Second War, much still remained. In France, for example, ex-Prussian State Railway G8 0-8-0s and P8 4-6-0s were running side by side with equally ancient indigenous designs and post-war 'liberation' 141R 'Mikados' in many areas into the 1960s. There is, therefore, plenty of equipment available for this project.

Before discussing the actual layout, a brief word or two about French railways by way of background would not go amiss.

France is a large country, considerably larger than the British Isles, so obviously its scenery, culture and railways are quite diverse. French railways were nationalised in 1938 to form SNCF (Société Nationale des Chemins de Fer Français). As with Britain, prior to nationalisation several separate railway companies served various areas. It should be noted, however, that unlike in Britain the planning of the whole route system was a governmental decision, and, on nationalisation, these companies formed the general basis of the six regions of the SNCF, in much the same way that British Railways had the London Midland Region, Western Region and so on. The regions were numbered 1 to 6 and the locomotives carried as part of their identification number the number of the region in which they worked.

Thus locomotive 2-231K-44 would be locomotive number 44 of type 231K of region 2. The 231 denoted that the locomotive had in British parlance a 'Pacific' 4-6-2 wheel arrangement, the 231 indicating the number of leading, driving and trailing *axles* rather than wheels. Thus, a 2-10-0 becomes a 150, and so on.

The six regions of SNCF are as follows:

1 Region de L'Est
2 Region du Nord
3 Region de L'Ouest
4 Region du Sud-Ouest
5 Region du Sud-Est
6 Region de la Méditerranée

The station of Clochmerle en Beaujolais. This well-known layout incorporates a number of features which can be related to our simple layout. The low continental platforms can be clearly seen as can the effective use and painting of a standard Jouef plastic station building kit. Note also the distinctive telegraph poles which are quite unlike those in the UK. The use of standard European locomotive models with slight modification, detailing and repainting is well illustrated. (Brian Monaghan, by courtesy of Ian Allan/Model Railway Constructor)

French branch line terminus 105

(Regions 5 and 6 were amalgamated in 1972 to form Region du Sud-Est et de la Méditerranée).

In addition to this main grouping of railways caused by the creation of SNCF, there were, and indeed still are, a number of other railways. Perhaps the best known of these are the CFD or 'Chemin de Fer Département' railways which had standard designs of track layout, buildings, locomotives and rolling-stock, and operated a number of systems across France.

There is plenty of information available on these and other small lines, both standard and narrow gauge, as well as SNCF, and the bibliography at the end of this chapter refers to some of these and other sources of information.

These secondary lines fall broadly into five types:
1. Lines leased to particular companies by the SNCF.
2. Lines classified as 'd'intérêt général' and operated by private companies but not under the SNCF umbrella.
3. Lines classified as 'd'intérêt local'. These lines have suffered closures fairly regularly from the 1930s.
4. Privately-owned industrial lines.
5. Tourist lines including lines broadly comparable to our preserved lines.

These railways were found throughout France and are particularly interesting and useful to the railway modeller because of the variety of locomotives and stock used, often bought second-hand or leased from the SNCF and other European companies such as Deutsch Bundesbahn.

The layout in this project will, then, depict a small rural 'd'intérêt général' line somewhere in France, and the purpose of the exercise is to provide a realistic, easy-to-construct layout which effectively portrays some of the flavour of the chosen area of the country and its railways circa 1960, when few of these lines remained but steam still reigned on those that did, albeit often in association with railcars.

Figure 65 shows the plan of the layout. The line could be anywhere in rural France that you choose, but remember that if you have a specific area in mind the style of architecture, building materials, trees and so on will vary. You don't actually have to know

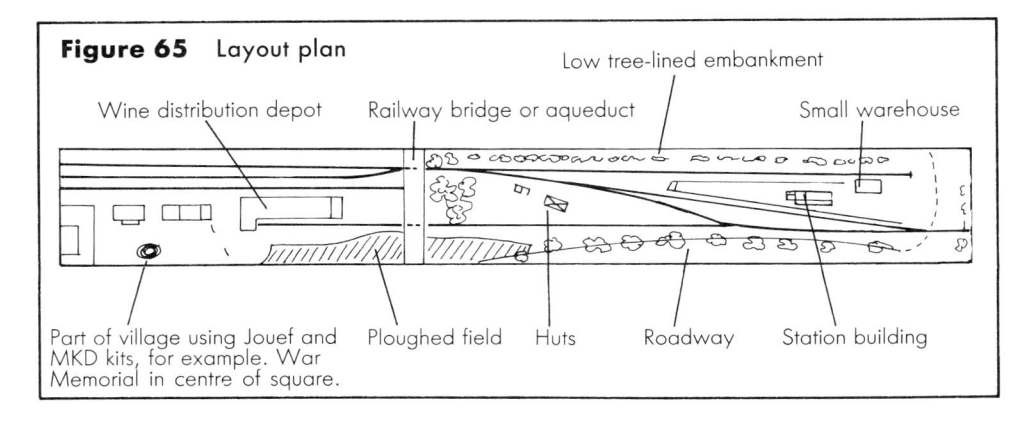

Figure 65 Layout plan

Wine distribution depot — Railway bridge or aqueduct — Low tree-lined embankment — Small warehouse

Part of village using Jouef and MKD kits, for example. War Memorial in centre of square. — Ploughed field — Huts — Roadway — Station building

Figure 66 Baseboard construction

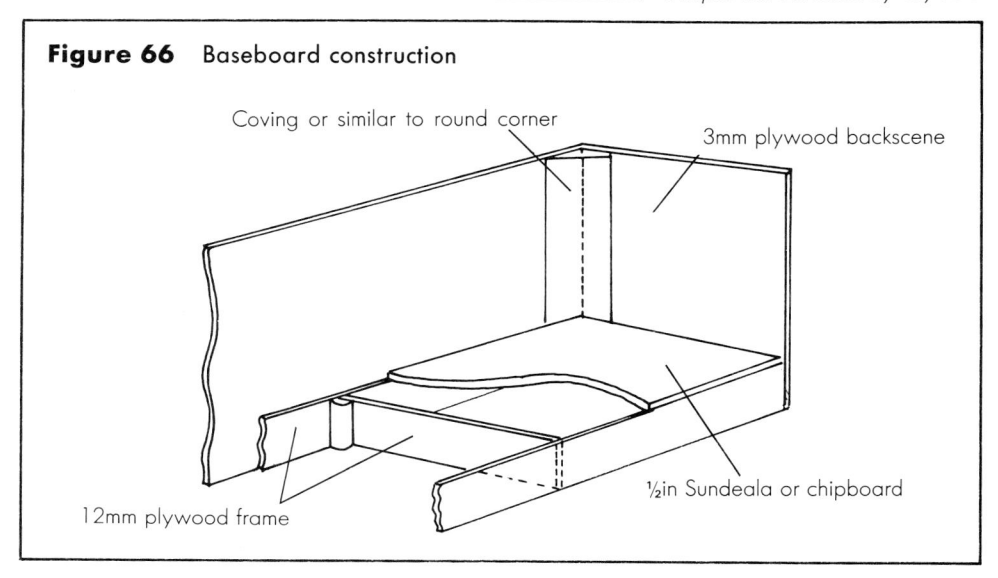

Coving or similar to round corner
3mm plywood backscene
½in Sundeala or chipboard
12mm plywood frame

the area well, as there are now plenty of large-format colour albums and books on rural France available from the library. Use the pictures to give the layout a definite character which reveals to the viewer its location and period even when locomotives and rolling-stock are not present. Also, try to marry the area of the layout to the reason for the railway's existence. It will probably serve a rural and largely agricultural community and maybe there is also wine production in the area.

Being set in about 1960, the station is likely to have a regular but limited passenger service to its main line junction. It is also possible, but unlikely, that a through coach to Paris or the regional capital might be attached to certain local trains and this would add to the operating interest. The addition of a small postal vehicle rather than a through coach would perhaps be more likely.

Freight operation would consist of coal, general merchandise, agricultural produce, livestock and the local wine. The livestock and general merchandise would be handled at the depot close to the station, and the wine from a private siding serving the distribution depot of the local producers' co-operative.

Baseboard construction

The baseboard is constructed in two pieces, each 4ft × 1ft, and Figure 66 shows the construction method. The outer frame is 12mm ply with 12mm ply cross members. For the longer side members it is worthwhile paying a little extra to have them machine cut at the wood-yard, emphasising, of course, the need for squareness and accuracy. All sub-frame joints should be pinned and glued. For ease of transport or storage, the baseboards are arranged to be bolted face to face but kept a little distance apart to protect the scenery and structures. Two end frames are used for this purpose, and Figure 67 illustrates how this is done.

The baseboards need to be strong, rigid and as light as is practicable. The top surface is Sundeala board; some of the larger model railway shops stock 4ft × 2ft pieces specially packed for railway modellers. There are other similar types of board, and one national DIY supermarket group stocks it both in full sheets and smaller part sheets which are ideal for model railways.

The baseboard is the fundamental element of a model railway and, whilst the construction may be a chore, if the frame is not square, the surfaces not flat and the joints between baseboards not accurate then it will reflect on the success of the whole layout.

The baseboards are joined together by hinges, preferably brass; the pin heads are filed off and the pins removed to be replaced by a length of a suitable diameter brass rod, bent to make removal easier. It is essential that the replacement pin is a close fit in the hinge (Figure 68).

Figure 67 Storage and carriage of baseboard using bolted end frames

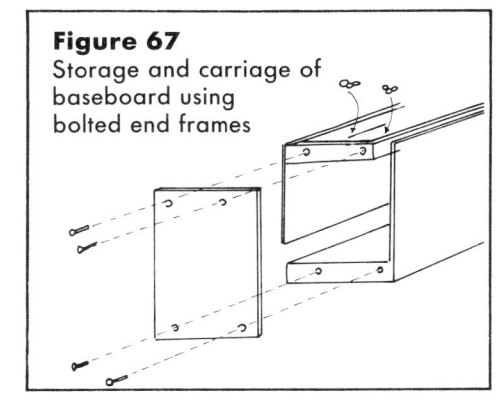

Figure 68 Joining the baseboards

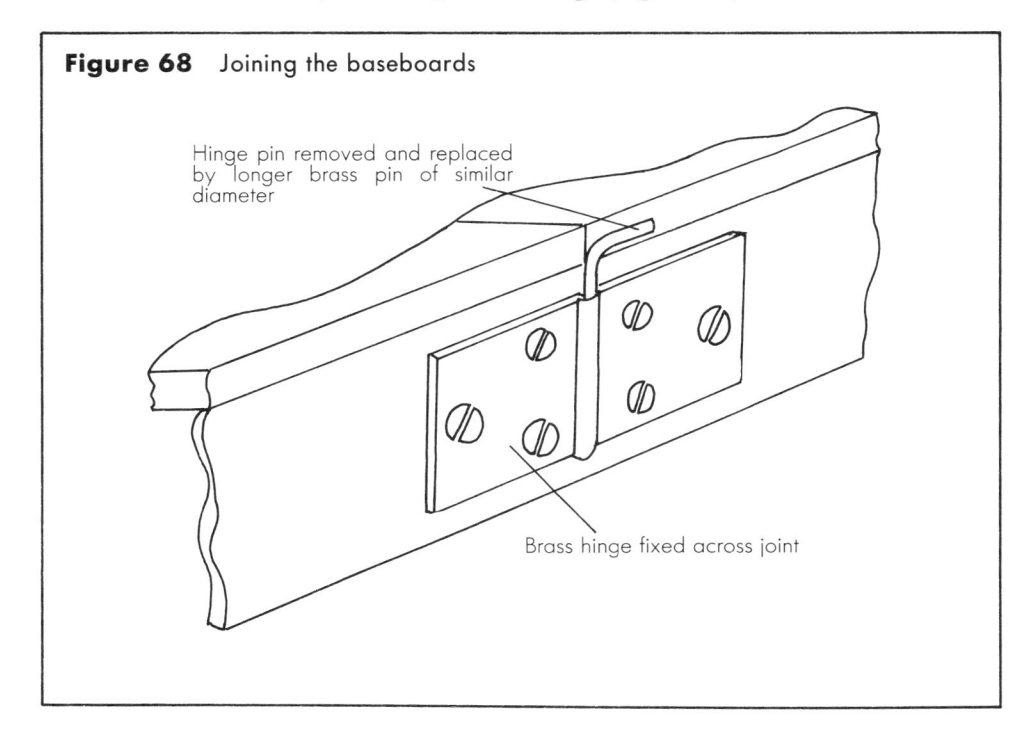

Hinge pin removed and replaced by longer brass pin of similar diameter

Brass hinge fixed across joint

Trackwork

To effectively represent the trackwork of the prototype on this model will probably require some compromise. It would in reality probably be constructed of flat-bottom rail of a fairly light section. Here we have a problem because the HO flat-bottom rail trackwork currently available from the flexitrack ranges is rather coarse and overscale, representing at best more modern heavy-duty rail used for high-speed main lines. The choice is between accepting a compromise and using trackwork from, say, the Peco Streamline range with live-frog points, or using a finer section rail to build a more accurately scaled trackwork yourself. The latter approach should not be dismissed as in my experience it is not difficult, and this extra effort is recommended and well rewarded by the appearance and running qualities of the resulting trackwork. Peco have recently introduced a system of self-build trackwork which should make this a much easier proposition. The SMP 00 gauge trackwork system which includes plastic-sleepered point kits is a very near scale representation of some of the older type of French trackwork and would provide an attractive alternative.

When laying flexitrack, as always ensure smooth changes in direction of the track on curves, avoiding sudden and severe kinks and bumps which may later give running problems. Also ensure that rail joints are carefully prepared and smooth, and avoid leaving burrs on cut rail ends. Where the track is joined and sleepers are cut away to allow room for the fishplates, cut the chairs from the affected sleepers and slide them back into the join to avoid unsightly gaps from missing sleepers. Finally, ensure that the fishplates are a tight fit.

Test the layout thoroughly at this stage. If a commercial flexitrack is used, careful ballasting can help to improve its appearance; moulded foam underlays should not be used for this layout as they represent the more modern tracklaying methods. Ballast (after wiring the layout) with the Woodland Scenics granite ballast; it is slightly more expensive than some other manufacturers' products but the fineness of grade and choice of colour make it ideal for this layout. It should be applied dry over the track and lightly brushed into the desired position and shape with a large, soft paintbrush, ensuring that it is kept clear of point blades and switches. A diluted PVA glue to which a drop of washing-up liquid has been added should be gently applied to the ballast with an eye-dropper or similar and left to dry overnight. Any surplus ballast can be lightly brushed away and collected for use again, and bald spots can be re-treated similarly. Remember that secondary lines tended to have sleepers totally buried in ballast and ash with just the ends showing if at all.

Wiring

It is assumed that if a commercial flexitrack has been used, the point motor systems which go with it will also have been used and mounted beneath the baseboard. Wiring of the point motors should, of course, be carried out according to the manufacturer's instructions and this is a straightforward operation.

A suggested wiring scheme is shown

Figure 69 Wiring diagram

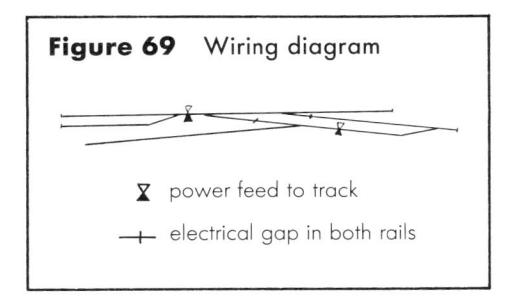

⊻ power feed to track
—|— electrical gap in both rails

in Figure 69. Electrical feeds to the track are shown and a number of section switches are required. These, together with the point motor switches and controller, should be mounted on a small removeable control panel which simply bolts on to the layout or is attached with loose-pin hinges. All wires should be fed to a multipin plug and socket at the edge of the baseboard, then harnessed together and neatly tied and fastened to the framework to prevent accidental damage. The use of multi-coloured wire, a different colour for each purpose, and a carefully-kept record of what each wire is for and where it goes, will help with fault-finding later. Ensure that the plug lead from the control panel is sufficiently long to enable the panel to be mounted in front of the layout for normal home use or mounted on the back for exhibition use.

Scenery

Attention can be turned to the scenic work once the layout works satisfactorily. Careful preparation of the backscene and its integration with the other scenic effects will be important both in completing the picture it is wished to create and in providing a frame for the layout.

The first task with the backscene is to disguise the corners, as the real sky does not have any! This is easily solved with the aid of expanded polystyrene coving as shown in Figure 66. Alternatively, cardboard with the edges suitable chamfered would do the trick. The backscene can be either covered in sky paper or painted a pale blue/grey.

The basic scenery is simple to construct using expanded polystyrene (ceiling tiles are ideal), Polyfilla, watercolour powder paint in black, red and yellow and Woodland Scenics foliage and turf materials.

The first and perhaps most identifiably French feature is the tree-lined roadway. There are a number of excellent ready-made poplar trees of the wire and flock type on the market which, although a little more expensive than plastic ones, are far superior. However, not all roads have avenues of poplars; different species such as lime and plane trees border the roads in other areas of France. Again, there is a wide variety available and it is worthwhile spending a bit of time making detailed trees for the few required or paying a little extra for the super-detailed models available from the specialist manufacturer. The area beneath and between the poplars needs some undergrowth and small bushes which can be easily represented using turf mix, rubberised horse-hair and real twigs. A few poppies made from fine brush hair topped with a blob of bright red paint would be appropriate if the layout is set in northern France. Finally, Figure 70 suggests how the cultivated fields adjacent to the road could be represented.

The roadway can be constructed

Figure 70 Ploughed field

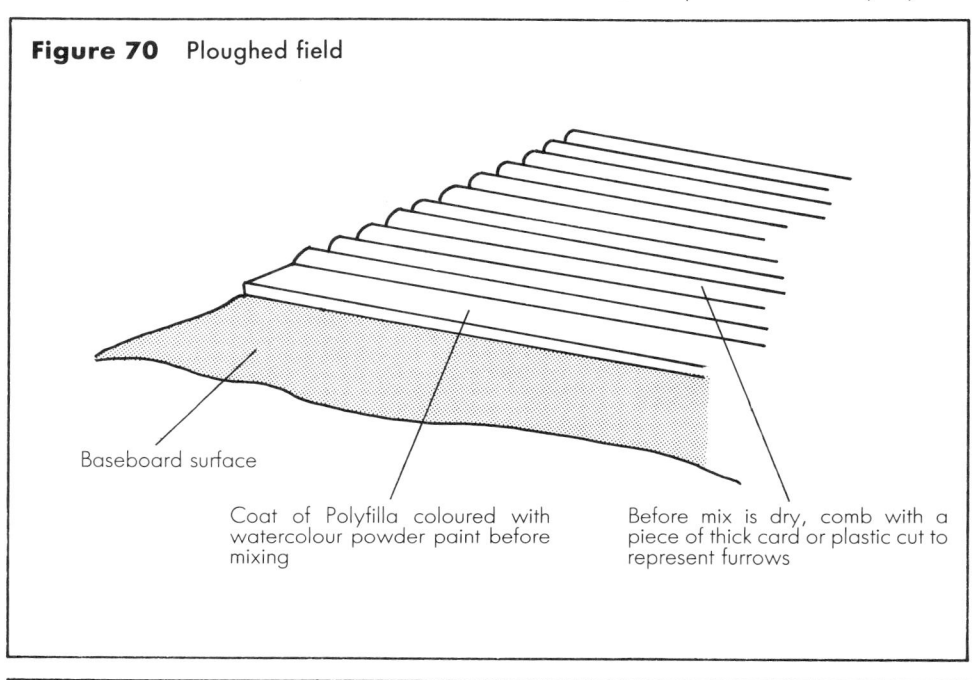

Baseboard surface

Coat of Polyfilla coloured with watercolour powder paint before mixing

Before mix is dry, comb with a piece of thick card or plastic cut to represent furrows

Figure 71 Roadway at front of layout

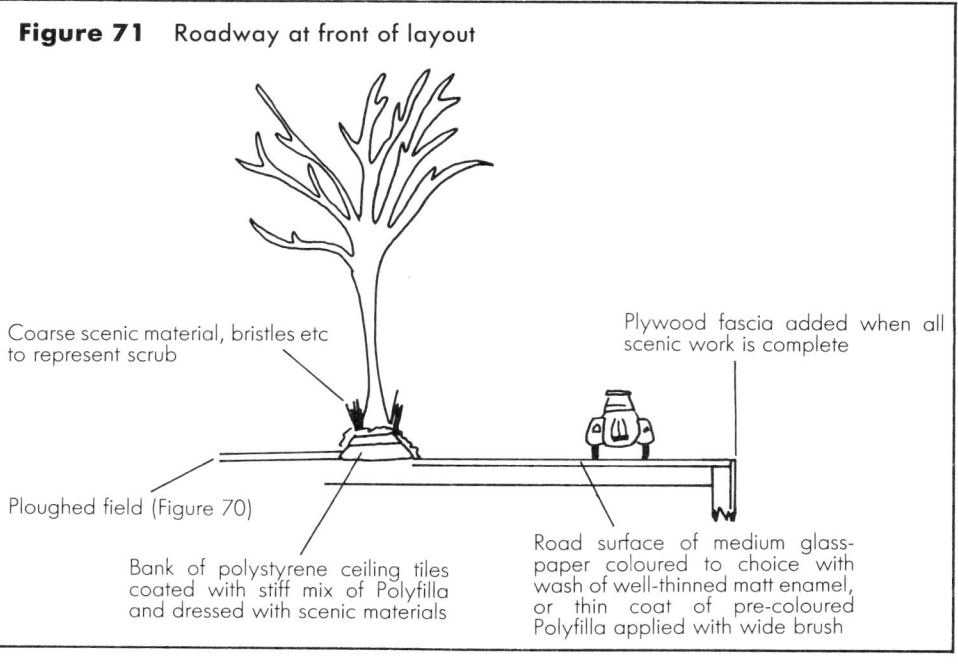

Coarse scenic material, bristles etc to represent scrub

Plywood fascia added when all scenic work is complete

Ploughed field (Figure 70)

Bank of polystyrene ceiling tiles coated with stiff mix of Polyfilla and dressed with scenic materials

Road surface of medium glass-paper coloured to choice with wash of well-thinned matt enamel, or thin coat of pre-coloured Polyfilla applied with wide brush

French branch line terminus

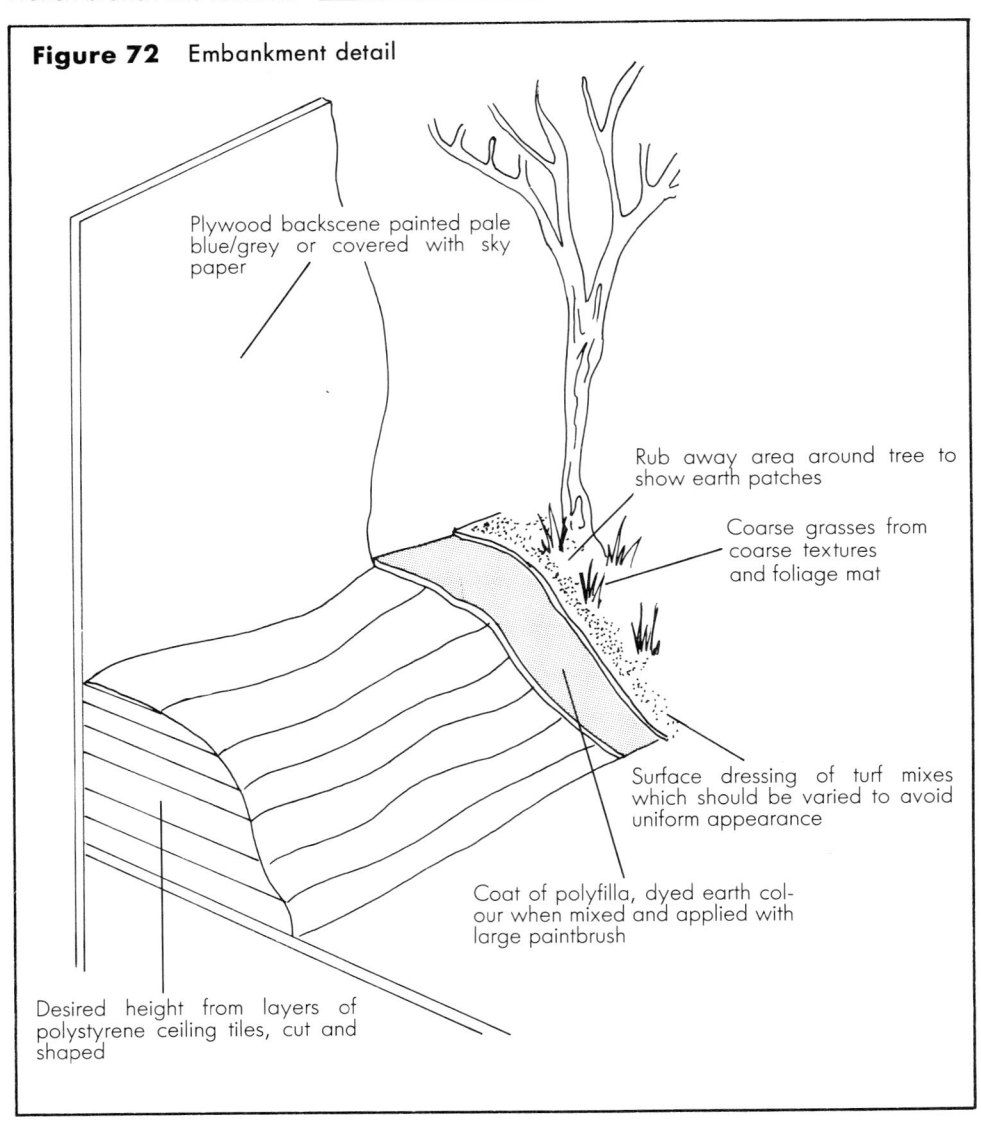

Figure 72 Embankment detail

- Plywood backscene painted pale blue/grey or covered with sky paper
- Rub away area around tree to show earth patches
- Coarse grasses from coarse textures and foliage mat
- Surface dressing of turf mixes which should be varied to avoid uniform appearance
- Coat of polyfilla, dyed earth colour when mixed and applied with large paintbrush
- Desired height from layers of polystyrene ceiling tiles, cut and shaped

from a very thin, even coat of Polyfilla, dyed a buff colour with powder paint and applied with a large, say 1½in brush. See Figures 71 and 72 for general guidelines.

The layout also requires a number of railway buildings. The basis of these are plastic kits from the Jouef and MKD ranges such as the Jouef 'Gare de Villeneuve' station building and the 'Gare et Quai de Marchandise' goods and wine depot. With a little painting, weathering and detailing, their source becomes less obvious and they take on

Simple model railway layouts

a welcome individuality. For example, the *station building* could receive an overall coat of a matt pastel colour on the rendering perhaps a shade or two darker than the stonework. Consider adding louvres to the upper windows, and change the slate roof to a pantile roof using the pantile sheets from the Wills scenic range painted a matt shade of orange/brick red with odd tiles picked out in other shades and dry-brush with green to represent moss. Many real French stations have buildings with ivy or some similar plant covering a wall or two and this is easily represented using a suitable shred of Woodland Scenics or similar foliage, thinned and teased out more the higher up the wall it goes.

In France, as in the rest of Europe, the station *platforms* are, by our standards, very low and can be easily

Above *The basic Jouef 'Gare de Villeneuve' building assembled as intended.*

Below *The Jouef 'Gare de Villeneuve' building shown in the previous picture after detailing and weathering. It is very useful for small layouts and is capable of a considerable amount of adaptation, shown here on a layout very similar to the one described. It is without a made up platform, this facility consisting merely of compacted earth with perhaps an ash or stone surface. The clutter and detail, typical of a rural station, is worthy of note, as is the ploughed field constructed as described in the text but with the addition of a coating of a fine brown-coloured sawdust dressing.*

French branch line terminus

Figure 73 Platform construction

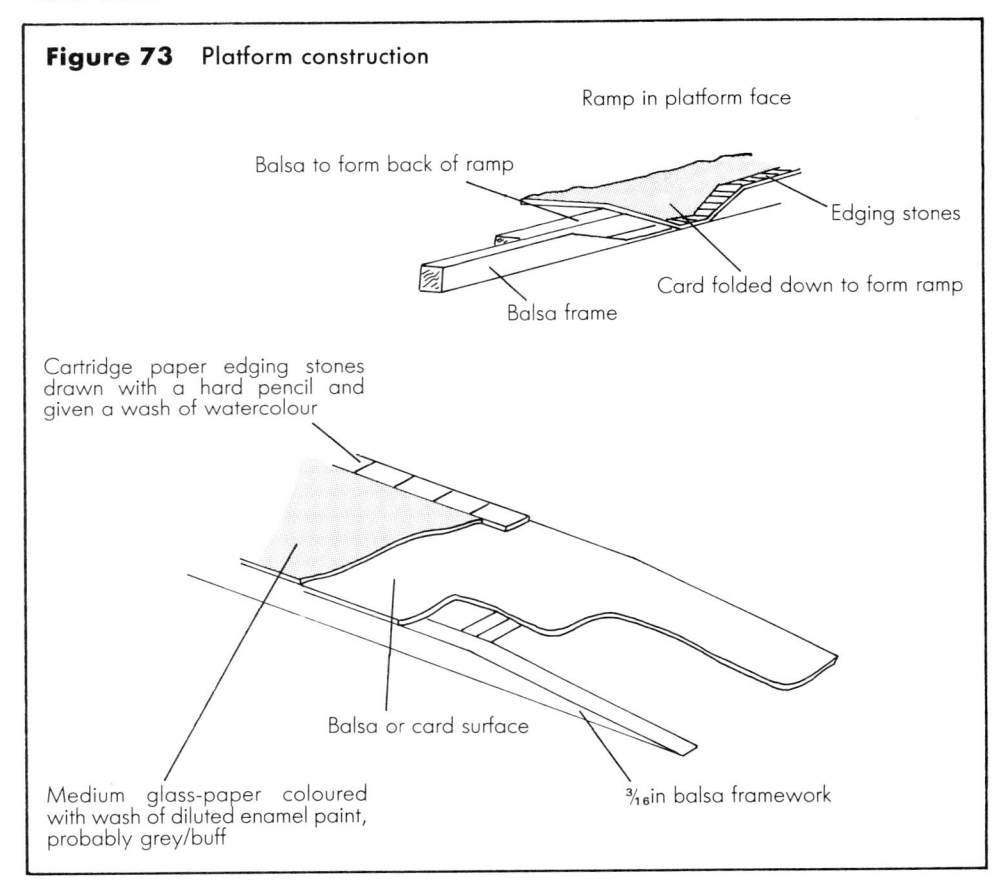

- Ramp in platform face
- Balsa to form back of ramp
- Edging stones
- Card folded down to form ramp
- Balsa frame
- Cartridge paper edging stones drawn with a hard pencil and given a wash of watercolour
- Balsa or card surface
- Medium glass-paper coloured with wash of diluted enamel paint, probably grey/buff
- 3/16in balsa framework

represented by a 3/16 in square balsa framework with a thin card surface. The actual platform surface finish comes from a further layer of medium grade glass-paper, suitably painted and with cardboard for the stone edges (Figure 73). (In a very small remote station, the platform would be no more than beaten earth raised to rail level.)

Finishing touches can be added ad infinitum but should include telegraph poles of the typical French design produced by MKD, figures from the Preiser or Merten ranges and a Citroen 2CV6 or Dyanne from one of the range of 1:87 scale road vehicles.

If it is wished to add more buildings to represent parts of the village in place of the countryside, both Jouef and MKD produce nice terraces of shops and houses and Jouef a school ('Ecole'). The choice is yours and the scene can be widened even further if you combine the various kits available.

If the layout is set in a more southerly part of France, the split between the station and fiddle yard could be achieved by providing an aqueduct or perhaps more typically a main line railway carried on a bridge. Figure 74

The road on this layout has been built as described in the text with the addition of a coating of fine surface dressing. The road vehicles help to date the layout as much as the stock used on it. The stone buffers are typical of the type used in France, and they are sometimes constructed of concrete. The goods platform has been fabricated from the platforms supplied with the station building kit and needs to be covered, at least in part. Using corrugated or wooden panels from the Wills scenic range, for example, it would be easy to add a lean-to goods shed.

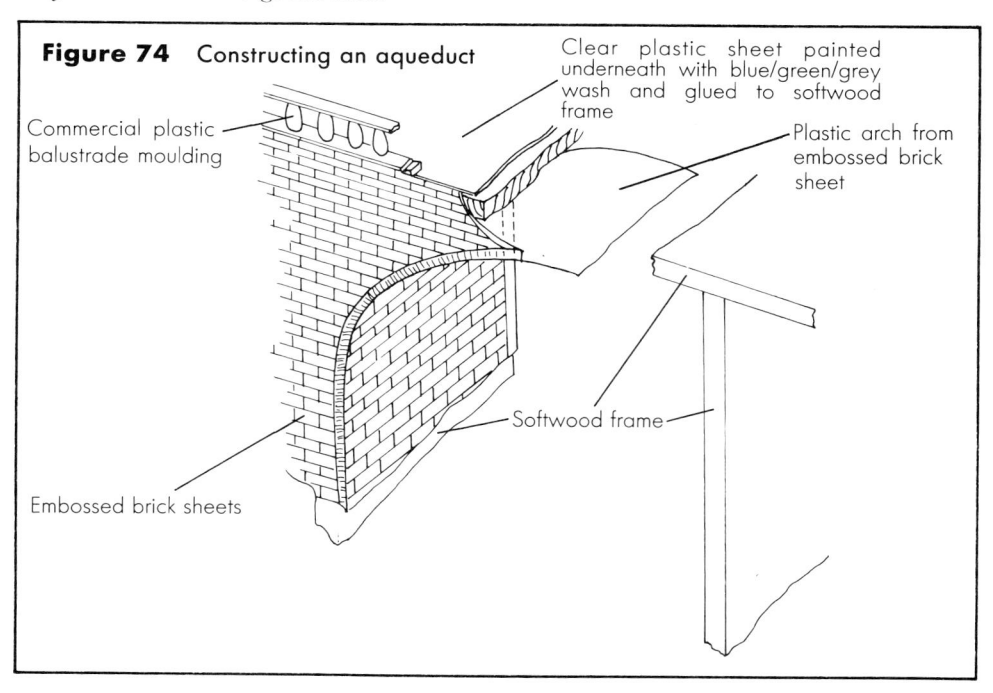

Figure 74 Constructing an aqueduct

French branch line terminus   115

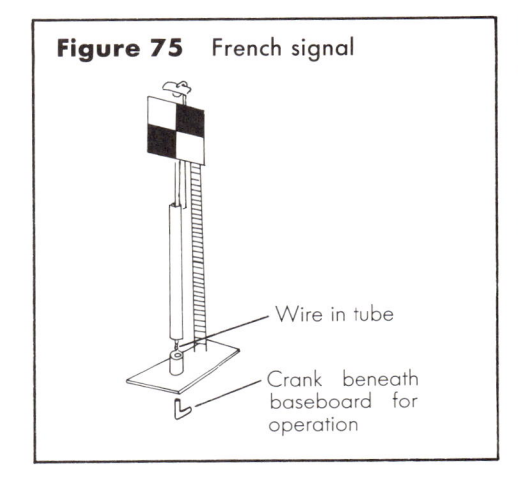

Figure 75 French signal

- Wire in tube
- Crank beneath baseboard for operation

shows how one may be constructed.

Signalling on French railways is very different from what we are used to in the UK. For purely scenic purposes, the layout could have a representation of the well-known chequerboard signal to help add the Gallic flavour. Figure 75 illustrates the construction, and suitable kits are available from MKD.

Locomotives and rolling-stock

Suitable locomotives and rolling-stock for the line can be chosen from a wide range which can be broadened considerably because the layout is a privately-owned 'd'intérêt général' line and has bought in stock from, perhaps, other main line railways including SNCF and may even also have leased or borrowed locomotives and stock. In this respect, the railway takes on a great similarity to the Colonel Stephens light railways in England.

There are many models of 'old-time', mostly ex-Prussian State Railway locomotives available, including antiquated tank locomotives such as the T3, and any of these delightful little models would add charm and character to the layout. Paint out their original markings and blacken the red wheel centres and underframes. Add new lettering from the rub-on letter transfers available from stationers to

Figure 76 Locomotive plates

The Jouef 140C is a suitable locomotive type for this layout. There are two types produced by Jouef, the 140C Ouest (Region 3) with small tender and the 140C Est (Region 1) illustrated. The latter has a large ex-German 34x tender and other significant differences from the Ouest version. (Peco Ltd)

represent the markings of your line. Figure 76 shows some typical designs.

As mentioned above, SNCF provided many of the locomotives and items of rolling-stock for these standard gauge lines towards their demise, and Jouef provide two very useful models of locomotive types which were often leased out. One is the 140C. These lightweight 2–8–0s were reliable and economical machines usually found in Regions 1 or 3 of the SNCF main line. In Region 1, they were coupled with ex-German type 34x tenders of high capacity whereas in Region 3 a smaller, rather quaint tender was used. Both variations are available from Jouef and each model incorporates various other detail differences. The other is the Region 3 0–4–0D. These tank locomotives were leased to various independent companies and similar

The detail on this ready-to-run Roco loco, a 050B, ex Prussian State Railways and later PLM, is quite amazing. It is supplied with clip-on number plates for PLM and later SNCF periods, together with a number of other clip on-details such as vacuum pipes, handrails and dummy scale couplings of commendable fineness.

French branch line terminus

The FNC railcar would be particularly useful for this layout, being only a short vehicle. The development of these vehicles is interesting, being a collaborative effort between Rail Unions and SNCF. The X5600 version is available in kit form from K's (N. & K.C. Keyser Ltd)

models were bought by the companies from the manufacturers who supplied them to SNCF.

There is, therefore, an excellent choice of locomotives available and, of course, locomotives leased to the companies from SNCF often kept their SNCF markings, so a repaint or re-lettering can be avoided if it is too daunting a task.

Railcars were common on French Railways from the 1930s as part of the effort towards greater economy of operation. There is a variety of mainly Renault-type railcars available through specialist UK importers from the firm of AS and these are even available in pre-SNCF livery. The British manufacturers N. & K.C. Keyser produce easy-to-assemble white metal kits for both Renault and FNC railcars which are supplied complete with motors and wheels.

There are also a number of small diesel shunting locos, or 'locotracteurs' in the ranges of all the continental manufacturers which would be equally suitable for this layout. It really depends on your preference for steam or diesel and, of course, the period

The French railways, in search of greater economy of operation in the 1930s, put a lot of effort into developing railcars, often in association with outside manufacturers. The British firm of kit manufacturers, K's, produce a number of kits for interesting European locomotives and two French railcars. Illustrated here is a Renault railcar. (N. & K.C. Keyser Ltd)

The Roco Y 8000 'locotracteur' is a very attractive little model and is shown in the latest yellow and brown livery. It is a modern prototype, however, and therefore only relevant to modern image layouts.

modelled.

A vast array of coach models from the various manufacturers is available. Older four-wheel and six-wheel stock meets the needs and only one or two vehicles would be required, perhaps as an alternative to a railcar. The six-wheel clerestory coaches from the Roco range, for instance, would be very suitable for the layout. A suitable 'through coach' would be one of the French-built coaches dating from the late 1920s found in the Roco range. Some departmental or postal vehicles would be a useful addition and again several choices are available.

Six-wheel third class coach from Roco.

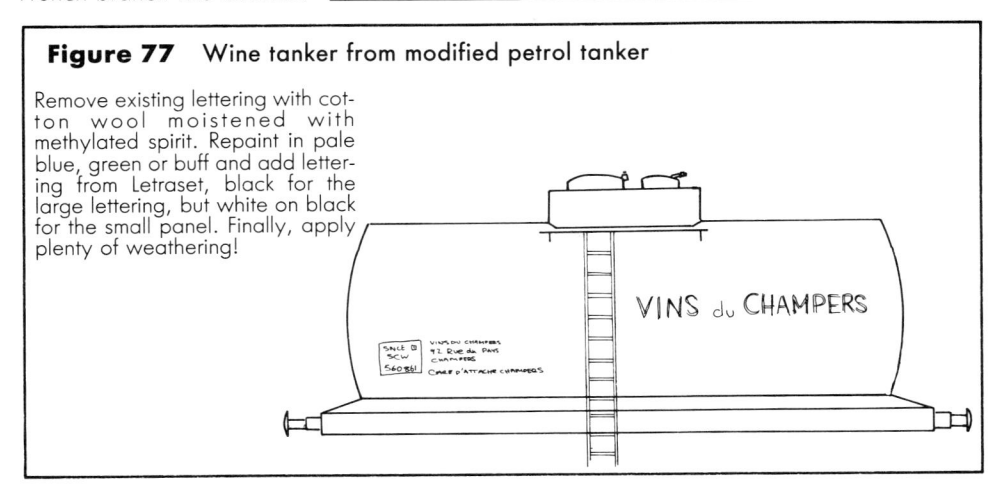

Figure 77 Wine tanker from modified petrol tanker

Remove existing lettering with cotton wool moistened with methylated spirit. Repaint in pale blue, green or buff and add lettering from Letraset, black for the large lettering, but white on black for the small panel. Finally, apply plenty of weathering!

All the major manufacturers of European systems produce a vast array of suitable goods stock. The stock for the layout should include a preponderance of the older, flat-roofed vans and wooden-bodied open wagons; avoid the more glamorous and colourful specialised vehicles. Some specialised vehicles will be required for the wine trade, however, such as two or three of the older wine cask 'bi-foudre'-type tanks. Repainting a couple of all-steel petrol tankers could also give an impression of the later all-metal winetanks. They could be lettered with Letraset or similar, and an example is given in Figure 77.

Both reference to prototype photographs of locomotives and stock and personal taste will decide whether the stock is run as bought or improved. Perhaps the simplest means of improving the already good-quality stock is to weather it; at the very least, the bright steel wheel trims, valve gear and in some cases handrails should be painted, the handrails matt black and the other items a mixed matt black and brown to represent the dirt and grease of a working steam locomotive (avoid getting paint on the wheel tyres and contacts!)

As far as operation is concerned, it is up to you, but short freight and passenger and even mixed trains would be the order of the day.

Further references

Magazines
Continental Modeller

Rail Miniature Flash, available in the UK through Peco

Loco Review, also available through Peco

Books
Steam on the SNCF, D. Bradford Barton

French Steam, Broncard and Felino, Ian Allan

French Minor Railways, W.J.K. Davies, David & Charles

CHAPTER 8
0 gauge light railway

Mention of 0 gauge in a book on simple model railways may well seem a little strange. However, it is very highly regarded and there cannot be many who have seen 0 gauge models on display and failed to appreciate its appeal. '0 gauge is very nice, but it's too costly and takes up too much space', is a comment often heard. These two common objections to 0 gauge model railways have some superficial truth in that individual items do usually cost more than similar items in the smaller scales, especially locomotives and stock. However, you require considerably fewer items to fill a given space in 0 gauge than in the smaller scales. For example, a small branch line layout such as that suggested in this chapter would in 4mm scale probably require four locomotives, six coaches and passenger-rated vans, and twenty or so goods vehicles. The 0 gauge layout occupying the same space on the other hand would have difficulty in physically accommodating more than two locomotives, two bogie coaching vehicles and at most a dozen goods wagons. The 0 gauge stock will almost certainly be kit built and if prices of comparable standard kits and components for the locomotives and stock are compared for the examples quoted, then the total cost will not be much different.

As far as the second objection is concerned, that of space, the layout plan outlined shows that an 0 gauge layout *can* be built in a comparatively small space. The suggested design is a fairly basic affair and whilst a more complex layout could be accommodated in the space available it was felt important to avoid overcrowding the layout. When one also considers that in all likelihood the rolling-stock and buildings will require kit or scratch building, then time also plays an important part in the interests of getting something running fairly quickly, with completion being an achievable target.

In short, there is a great deal of satisfaction to be gained from a small 0 gauge layout, far more in my experience than from much larger layouts in smaller scales. A layout in 0 gauge with two locos and a small selection of stock, all highly detailed, might just turn out to be your cup of tea. As with all the layouts in this book, because they are small a lot more time can be spent in achieving a high standard and

All the decrepitude of a light railway is captured in this post-war picture on the Kent and East Sussex Railway. Of particular note is the appearance of the locomotive, an ex-LBSC Terrier, and the ex-LSWR coach, which looks quite filthy. The weed-grown track in the foreground could easily be represented on the model by adding coarse scenic dressing material to the ballast in the sidings. It would also be quite easy to adorn a model locomotive with a bucket, fire-irons and even the fireman's bike on the footplate! Excellent detailing parts for this type of work are available from Springside. (Author's collection)

level of detail, and detailing, of course, is much easier in 0 gauge!

The plan

The idea behind the layout is to represent a very basic branch line terminus in a rural part of southern England. It is almost a light railway, providing only basic facilities with definitely only 'one engine in steam' operation. There are, or rather were, real-life examples of stations such as this — Loch Tay in

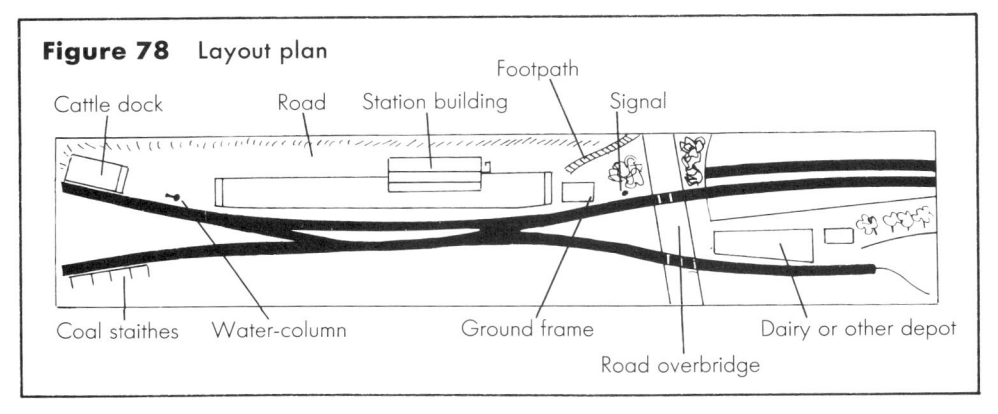

Figure 78 Layout plan

Scotland and Highworth in Wiltshire (Caledonian and GWR respectively) spring to mind. This layout, however, is set in south-east England and while it is closely akin to the Kent and East Sussex Railway, with a change in the style of buildings and perhaps a few less trees and a less lush scenic dressing, any area could be represented.

The layout plan (Figure 78) provides for a short platform, cattle dock, general goods siding and, of course, a run-round loop. There is also a private siding and this can serve the industry of your choice, perhaps a fruit packer, dairy or milk depot. If you have a specific area in mind, the choice of industry will follow naturally; for example, a dairy would lend itself to a West Country setting.

On a minimum-space layout such as this, the length of trains is going to be limited to one bogie coach or four or five wagons and a small locomotive, probably a tank engine. The lengths of the platform, run-round loop and fiddle yard reflect this.

Baseboard construction

The layout will occupy an area of 10ft 6in × 2ft and is carried on three baseboards each 3ft 6in × 2ft. The width of the layout could be reduced by a few inches but would then require reductions and modifications to the suggested scenic treatment.

The baseboards are easily constructed from a 2in × 1in softwood frame topped with ½in chipboard, as shown in Figure 79.

Trackwork

There are a number of alternatives for the trackwork. Peco produce an excellent range of 0 gauge track and pointwork while Alan Gibson pro-

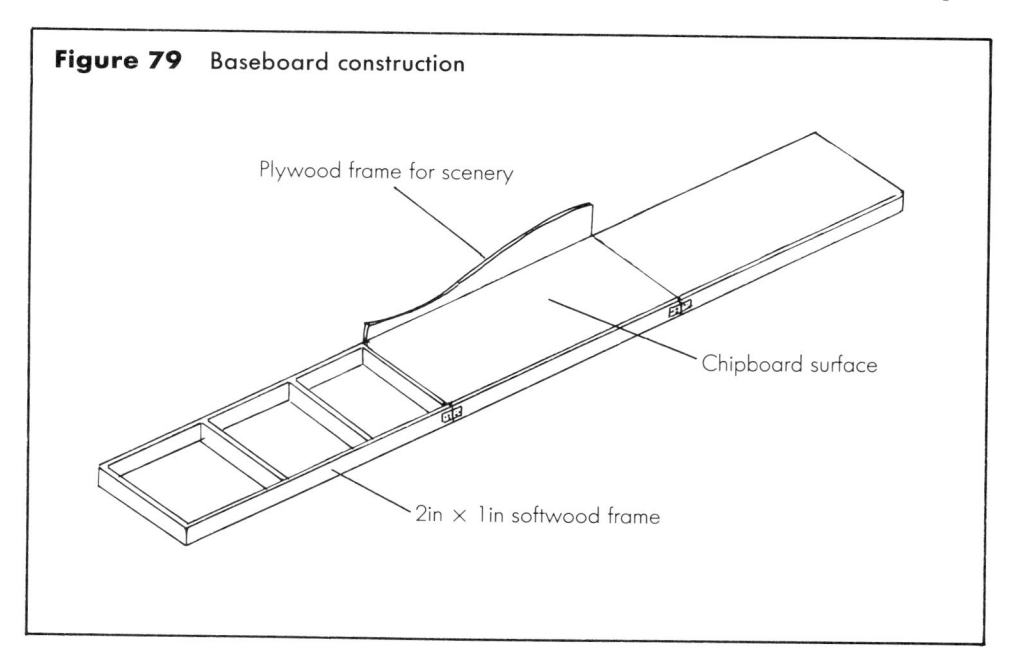

Figure 79 Baseboard construction

Plywood frame for scenery
Chipboard surface
2in × 1in softwood frame

O gauge light railway

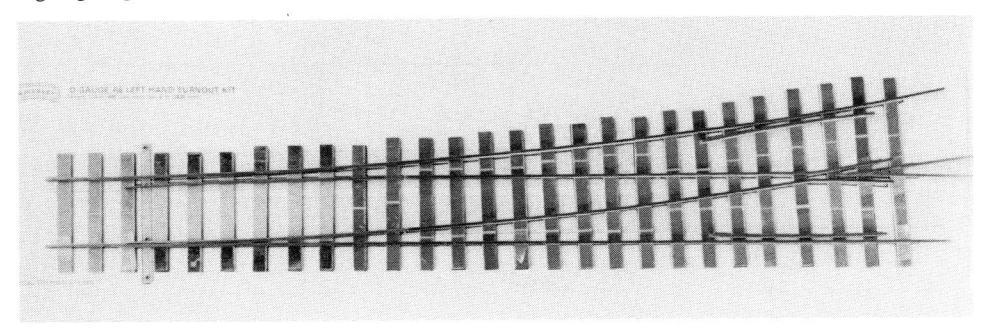

Construction of pointwork using the tried and tested method of soldering rail to printed circuit board or copper-clad sleepers is relatively easy with well-designed turnout kits such as those in the Waverley range. Wiring is simple, and requires only a two-way single pole changeover switch connecting the crossing switch to either polarity in accordance with the road selected.

duces superb 0 gauge flexible track and a range of components for point construction. There is also a range of easy-to-assemble point kits using rail soldered to copper-clad paxolin sleepers sold under the name Waverley. They come complete with the rail and with crossing vees machined and assembled, point blades ready machined and sleepers cut and gapped and the finished result is illustrated above. They are excellent value for money and can be used with either of the two flexitrack systems mentioned. The notes which follow, however, assume the use of the Peco track and pointwork.

In an attempt to minimize noise, the trackwork is laid on an underlay, $\frac{1}{8}$in cork sheet being the traditional material. However, greater noise insulation can be obtained if a fairly hard grade foam rubber is used, and the type sold for basic carpet underlay is ideal.

Carefully mark the centre line of the track on the baseboard (full-size paper templates of the pointwork are available at a nominal charge from Peco and are a helpful aid when planning the layout). It is essential to take your time and ensure that adequate clearances in loops and sidings are provided. Once the track plan is finalised, the underlay material is cut to cover the baseboard where the track is to go and then laid to the marked centre line. Where there are parallel tracks, they are covered by a single piece of material, but for single track, such as the siding, use strips of material some $1\frac{1}{2}$in wide. Latex-type glue is ideal for fixing the underlay. The track is then glued down on to the underlay, but avoid getting glue around the points tie bar.

Wiring

The wiring of the layout is simple and requires three feeds, one at the entrance to the station and two to the run round loop; the wiring diagram (Figure 80) shows this. They are switched by simple on/off switches located in the control panel with the controller and

Figure 80 Wiring diagram

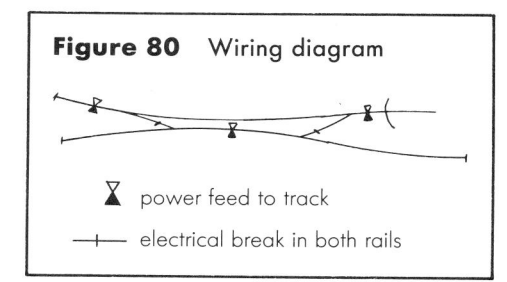

▼ power feed to track
—|— electrical break in both rails

the point switches.

Peco point motors can be used and while there are a number of options for installing them, most modellers place them beneath the baseboard. They should, of course, be installed and wired in accordance with the instructions supplied. The wiring should be collected in a harness and run via multi-pin plugs and sockets from baseboard to baseboard and then to the control panel.

Scenery

Once the track is fixed in position, the glue has thoroughly dried, the wiring has been installed and everything thoroughly tested, ballasting can take place.

Loose ballast, either cork or granite chips, is spread over the trackwork and brushed into the desired position. Once this is done and any details have been added such as ash in the areas where the locomotive may stand, or weeds and thicker piles of spent ballast around the buffer stops, a 50/50 mix of PVA glue and water is applied to the ballast with an eye-dropper or similar. A drop of washing-up liquid added to the mix will help. When this is thoroughly dry (it will need to be left for at least 24 hours) surplus ballast is brushed away; any 'bald' spots can then be re-treated and the track and ballast painted to taste using rust and track colour paints. Care is needed to keep the surface and inside upper face of the rails clean for electrical pick-up by the locomotive wheels.

There are not many railway buildings or structures on the layout and the construction principles outlined in earlier chapters for making buildings from plastic sheet apply equally well in 7mm scale. Thicker sheet materials will be required for the basic shells, and embossed brick and stone sheets, brick papers and fittings such as chimney pots are available for 0 gauge. Indeed, a great many 00 gauge components can be used to good effect in the larger scales, such as pantile roof sheets and cast details.

The place to begin the scenic treatment is with the *road overbridge* which hides the fiddle yard entrance and provides a scenic break. Peco produce an excellent plastic road overbridge with simple bridge sides and retaining walls. Because it will be placed at the edge of a baseboard and may be in a vulnerable position if the baseboard is moved, it is advisable to build the bridge around a simple hardboard or plywood box to provide not only the necessary strength but also to enable the road approach to the bridge to be extended to both edges of the baseboard.

It is extremely unlikely that a road bridge would in reality have been built over such an insignificant line; a level crossing would be more in keeping, but would not, of course, provide the scenic break. It could be explained that the railway was intended to expand and that the bridge was built in anticipation of development which, as was often the case, never materialised.

0 gauge light railway _____ 125

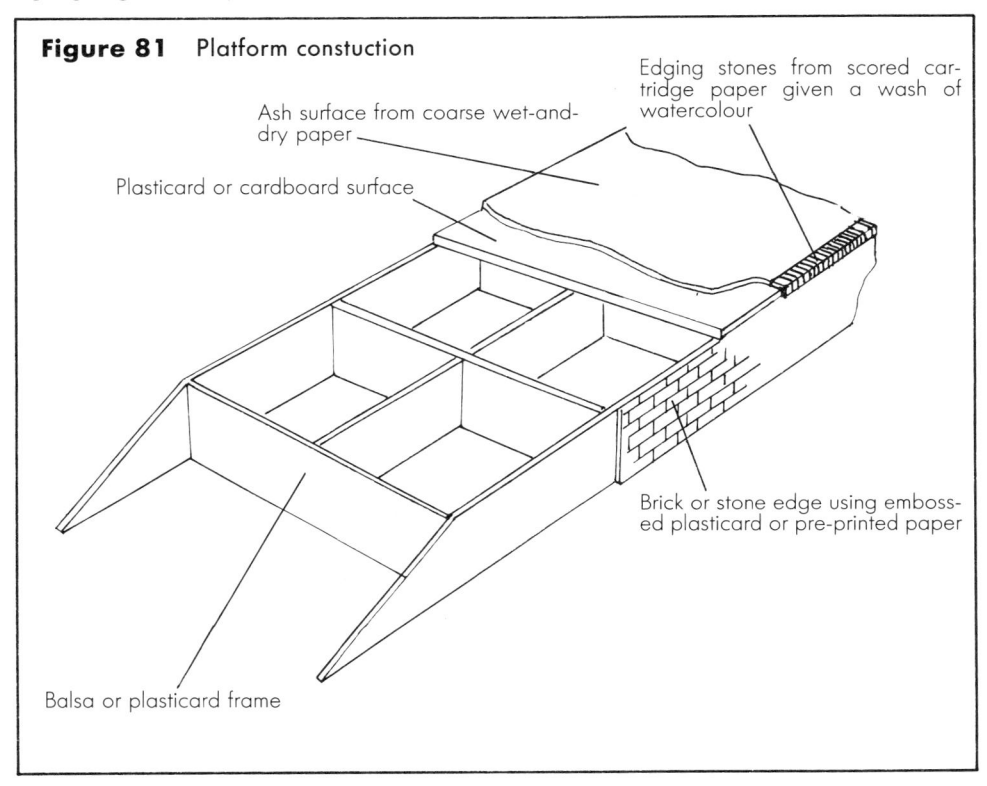

Figure 81 Platform constuction

- Edging stones from scored cartridge paper given a wash of watercolour
- Ash surface from coarse wet-and-dry paper
- Plasticard or cardboard surface
- Brick or stone edge using embossed plasticard or pre-printed paper
- Balsa or plasticard frame

If the layout is set in the south-east, perhaps Kent or Sussex, the bridge will probably be constructed of brick. It would not be too difficult to convert the Peco stone bridge to brick by covering it in embossed brick sheet or even brick paper.

The next structure to be made is the *platform*. This is constructed much as described in previous layouts, but forty thou plasticard or ⅛in balsa provides the frame. The photograph of Northiam station shows the type of platform and station building envisaged, but rather than leave the platform edge as at Northiam, you may prefer to use edging stone. Ash would in all probability be the surface dressing and this can be effectively represented by a very coarse wet-and-dry paper or alternatively coarse glass-paper which can be painted with washes of matt enamel paint to represent ash. Figure 81 shows the construction.

The *station building* suggested (Figure 82) is a 'tin shack' of the type favoured by Colonel Stephens on lines for which he was the engineer. They can still be seen in their basic form at Bodiam and Northiam on the Kent and East Sussex Railway and in their more refined form at Tenterden Town, the headquarters of the line. Similar structures appeared on many other lines including the Hawkhurst branch and the Sheppey Light Railway.

Northiam station on the Kent and East Sussex Railway is a typical Colonel Stephens 'tin shack', as shown in Figure 82. The platform, looking somewhat the worse for wear, shows the ash surface and brick edging quite clearly.

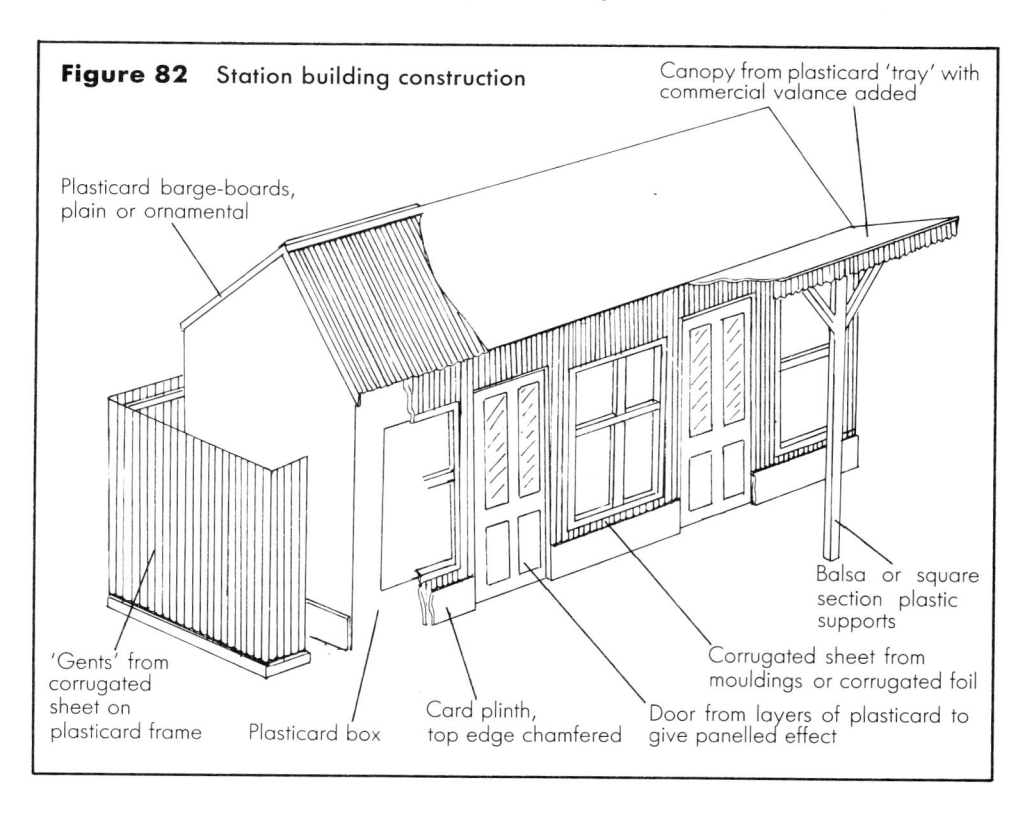

Figure 82 Station building construction

Canopy from plasticard 'tray' with commercial valance added

Plasticard barge-boards, plain or ornamental

Balsa or square section plastic supports

'Gents' from corrugated sheet on plasticard frame

Plasticard box

Card plinth, top edge chamfered

Corrugated sheet from mouldings or corrugated foil

Door from layers of plasticard to give panelled effect

0 gauge light railway

The prototype was constructed of corrugated metal sheeting or cladding fixed to a wooden frame, and this can be quite easily represented in model form. The basic shell is a box of forty thou plasticard with window and door apertures cut out, to which is added the cladding of corrugated iron sheet. A plastic moulding representing this material is available and whilst intended for 4mm scale models is equally effective in 7mm scale. Alternatively, corrugated alloy foil is sold in rolls for making Christmas decorations and this can be cut into equal-sized pieces, the corrugations running along the length, and then glued on to the shell, the upper sheets slightly overlapping the sheets below. Thin 'slices' of plastic rod can be glued on to represent the fixing bolts.

Doors and window-frames are added from plastic strip and section and a door built up from plasticard sheet. The roof is also made from sheets of this cladding. It must have been noisy during a hail-storm inside these buildings! The roof panels are glued on to the plasticard shell and the roof detail is finished with the addition of a ridge piece (Figure 83) and bargeboards on the ends. The latter can be either cut out ornamentally from plastic sheet or left as a simple strip.

The canopy is a simple plasticard assembly, as shown in Figure 82. Moulded or etched platform awning valance is available and can add an ornate touch to a simple building.

The building can be painted with enamel colours of your choice, depending on the livery of the company and the period in which the model is set. For example, if British Railways Southern Region is chosen, a

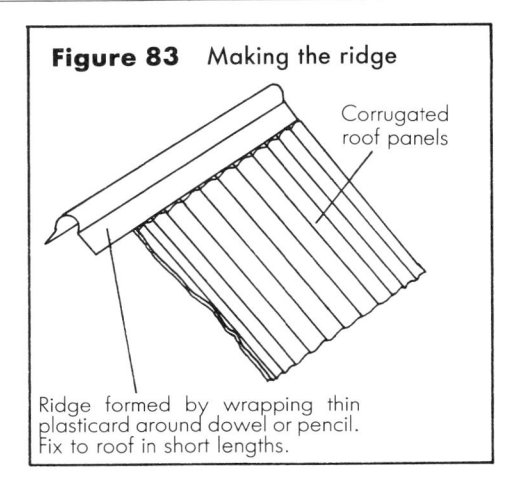

Figure 83 Making the ridge

Corrugated roof panels

Ridge formed by wrapping thin plasticard around dowel or pencil. Fix to roof in short lengths.

cream building with green doors and window frames would be typical.

The pens of the *cattle dock* are made from commercial plastic fencing suitably modified with additional posts and gates formed from plastic section and strip. The platform is constructed in the same way as the passenger platform but, instead of the ash surface, stone, brick or concrete is required.

Before the buildings are fixed to the layout, give any bare baseboard a coat of matt black or dark brown emulsion. The area at the front of the baseboard is flat and grassed, and this can be given some detail which will provide a visual draw to lessen the impact on the eye of the baseboard edge. The ground will need to rise towards the road overbridge and this is achieved by building up the contours with polystyrene ceiling tiles covered in a coat of Polyfilla or similar. This is later hidden by the fitting of a shaped $\frac{1}{8}$in plywood fascia to the outer edge of the baseboard when the scenic work is complete.

Care needs to be taken with the treatment of *grass and ground cover*, as in 7mm scale you can't really get away

Building up detail and texture in scenic work has a far greater significance in 0 gauge, where it is essential to show some separate growth. Pieces of coarse grass mat are shown here being added to a layout and make good representations of the undergrowth found on embankments. Separately-applied rushes are just visible in the bottom right-hand corner and are easily made from bristles. Further detail can be added from the etched plants available from the BTA and Scale Link ranges. (BTA Hobbies)

with pouring various shades of dried sawdust over glue. Dyed sawdust scenic dressings have their uses but, because of the increased size of everything, the opportunity to provide considerable detail, down to individual plants, should not be missed. This is actually a much easier proposition than it sounds, and there are some excellent commercial products available to help.

The basis of the grass is surgical lint which is glued down to the baseboard. The lint can be pulled away from the baseboard after the glue is dry to leave 'tufts' of grass, or could be combed and manipulated to achieve the desired effect. Colouring can be achieved by soaking the lint liberally with water-based colours, and detail and highlights can be added with designers' gouache or acrylic colours. Changes in texture can be achieved by using dyed sawdust glued on to the areas desired with a mix of 50/50 PVA adhesive and water. Better still, use the various turf and foliage dressings from the Woodland Scenics range; the finest turf in this scale is really only suitable to represent a fine lawn, but coarse turf could be used to represent fine 'sheep-grazed' turf. The foliage mats could be cut up into small pieces, rolled loosely and then teased out and glued in place to represent coarse plants.

The Scale Link range provides etched frets of reeds and tall grasses which represent such diverse species as thistles, poppies and wild cabbage. Separate frets provide bracken, dock and a variety of other common plants. These can be painted and flocked if necessary and planted in the grassed

0 gauge light railway _____ 129

area as desired. Usually these plants appear in clumps and the Scale Link frets are available in different scales so it is quite easy by mixing plants from a few different frets to avoid too much uniformity in appearance. The smaller bracken leaves can even be added to buildings around their base or even, trimmed down, growing from gutters and roofs. Look at the countryside itself or even some of the large format countryside album-type books in the library.

Further textures and details can be added by using short lengths of wool glued on end and teased out. Reeds or coarse grass can also be represented by planting bristles from old toothbrushes and the like and these, together with a liberal use of the coarse turf previously mentioned, can be quite effective. The rear of the layout can be similarly treated.

The goods yard area, such as it is, needs a surface and ash seems as likely as anything. Here again, recourse is made to the Woodland Scenics range for a suitable mix and also for a light brown fine-grade ballast which can be glued down to represent the gravel road to the station.

Trees can be made in a number of ways. The easiest is to use suitable twigs which can be planted in the baseboard and foliage added from the Woodland Scenics foliage mats. Alternatively lengths of braided wire sold for hanging pictures can be twisted together to form a frame of trunk and branches. The main sections of the trunk are coated in Das or similar and the foliage is added from the foliage

Construction of trees from etched parts is simple and, whilst not the cheapest method, the results certainly justify the slightly higher costs. Kits and parts are available from various sources, as are completed ready-to-paint models, at reasonable prices. (BTA Hobbies)

Figure 84 Constructing trees by the twisted wire method

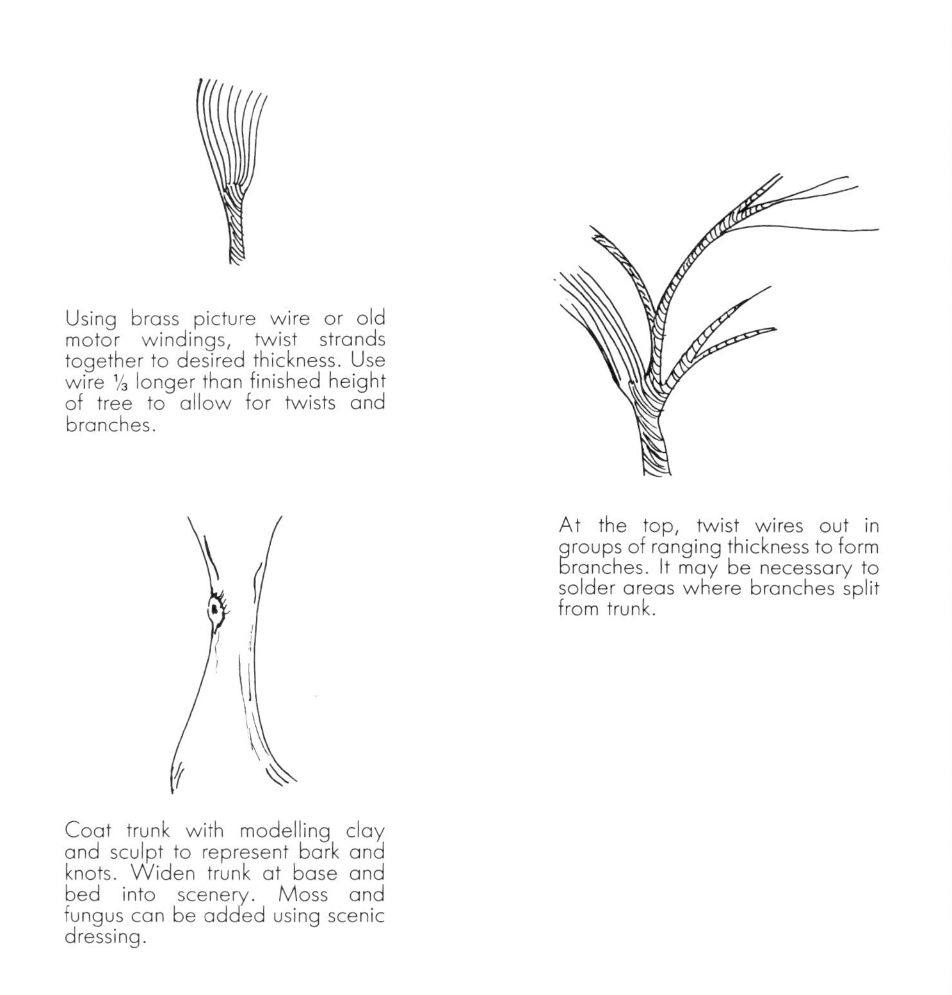

Using brass picture wire or old motor windings, twist strands together to desired thickness. Use wire ⅓ longer than finished height of tree to allow for twists and branches.

At the top, twist wires out in groups of ranging thickness to form branches. It may be necessary to solder areas where branches split from trunk.

Coat trunk with modelling clay and sculpt to represent bark and knots. Widen trunk at base and bed into scenery. Moss and fungus can be added using scenic dressing.

mat. Figure 84 shows the basic construction. The Scale Link range provides some exquisite cast branches and trunks to which etched leaves and fine branches are added. These components make the most realistic trees but are expensive, particularly if more than the odd one is required. If you make the trees yourself, use a simple book which shows various species in silhouette as it is the realistic outline of your model trees rather than leaf detail which will create the greatest effect. Small bushes can be represented by some of the small trees sold for N and 00 gauge, suitably roughened before

0 gauge light railway

The important thing in modelling trees is to get the overall shape right. Reference to children's tree identification books is an important aid to this. Note also the coarse undergrowth between the railway and the boundary fence. (BTA Hobbies)

planting to lose some of their uniformity. Fallen logs and tree trunks can be added from broken twigs and the ground beneath trees and bushes darkened and browned.

Fencing can be added immediately around the station, and post-and-wire components are available as are conventional post-and-rail and palisade fencing and gates.

Figures, animals and road vehicles (there are plenty of die-cast collectors' cars in 1:43 scale which can often be picked up quite cheaply) complete the scene. There are a whole host of barrows, packages, milk churns and the like, and a good many of those sold for 4mm scale will not look out of place on the 0 gauge layout.

Point rodding can be represented by 1mm square brass or nickel silver wire running on simple posts of ⅛in square balsa or plastic. If you incorporate a signal, signal wire made from 5 amp fuse wire can be run back to the ground frame from ⅛in posts planted every 1½in or so in the baseboard beside the track towards the outer edge of the track bed.

Locomotives and rolling-stock

0 gauge locomotives and rolling-stock can be bought ready-to-run, but since they will be kit or scratch built they will be very expensive. Lima produced a Class 33 diesel and LMS 4F locomotive in 0 gauge until quite recently, and these can still be found at reasonable prices second hand; the latter formed the base of some very interesting conversions to other types. The same firm also produced models of the BR MkI composite and brake second coach and standard 16-ton mineral wagons. These models can also still be obtained second hand and,

Fine-quality, simple-assembly wagon kits are readily available in 0 gauge and the range is steadily increasing. Three examples from Slaters range are shown, including a private-owner wagon with pre-printed lettering. These kits have very fine detail and sprung buffers.

whilst leaving something to be desired as to true scale, they can with a little effort be made into reasonable models.

There are plenty of easy-to-construct wagon kits available in 0 gauge, usually in plastic, covering a variety of types of vehicles, some of which are illustrated here. Plastic coach kits are starting to appear but, unless you can afford to buy kit-built vehicles, you are going to have to assemble them yourself from the wide variety of kits available. The majority are in metal with either aluminium or etched brass pre-formed bodies to which details and underframes are added. There are also some

The Southern Railway standard brake-van shown here under construction is an ABS white metal kit, and illustrates the fine detail to be found in the fast-growing ranges of 0 gauge kits and parts. These kits are easily assembled using glue.

0 gauge light railway 133

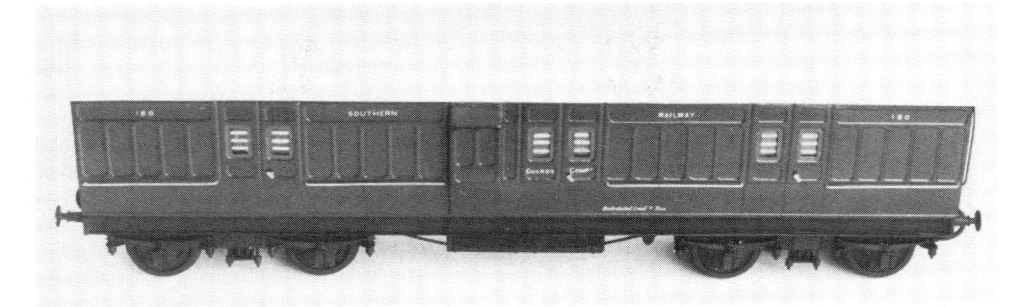

Rolling-stock kits in 7mm scale are not all metal; Highfield Models produce a number of plastic-bodied vehicles in kit form. They provide only the basic body parts which require detailing, but this is not difficult as there is an extensive range of coach and wagon parts, from cast door-handles to complete bogie kits.

very basic but quite cheap plastic coach body shell kits available which consist of formed plastic sides and ends with integral panelling; you will need to cut out windows, build a strong substructure and add underframes and bogies which are available separately. They can make pleasing models but do demand great care if they are to look good, particularly in cutting out the windows. As with all modelling, these kits require care and patience rather

Etched kits are becoming increasingly the norm for high-quality, detailed locomotives and rolling-stock. The better examples come with some parts pre-formed, as in the coach kit shown here. Though good results can be achieved using modern 'super glues', this is not always the best method.

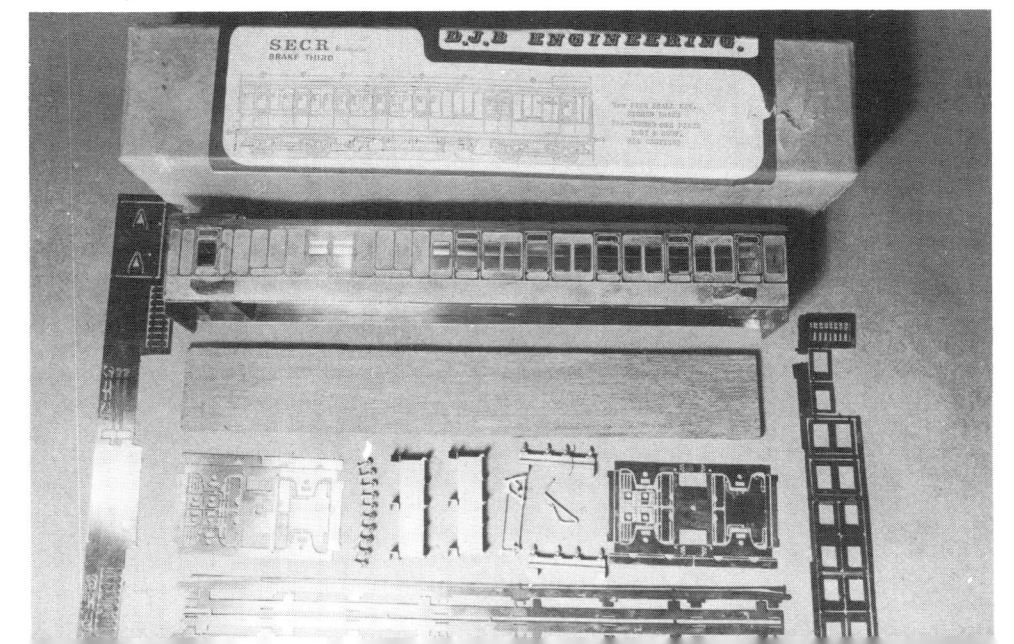

The recent resurgence of interest in 0 gauge has led to the introduction of new generations of kits and accessories, far superior in quality to earlier products. Typical of the latest in 7mm scale locomotive kits is this ex-LSWR B4 tank locomotive from Vulcan. It provides, as can clearly be seen, a super-detailed model which is not difficult to construct, even for a relative beginner. (Ron Cadman)

than great skill.

Unless you can either afford to buy ready-assembled locomotives or can find and adapt a Lima 4F to your choice of prototype, you are left with building your own kits. The recent revival of interest in 0 gauge has led to a large increase in the number and choice of locomotives available in kit form. Whilst many are in etched brass and designed for the more experienced modeller, there has been a trend recently towards very highly-detailed and well-designed locomotive kits utilising a cast white metal body and brass chassis. The body can be assembled with glue and the chassis usually bolts together or may require a minimum of soldering with a 'fold up' assembly. Kits falling into this category include the Springside GWR 14xx and the Vulcan Terrier, B4 and 04 diesel. They provide a great many parts to build a really well-detailed model and also contain good instructions which should guarantee success to all but the most inexperienced modeller. CCW also provide some basic kits with white metal bodies, and screw-assembly chassis kits are also available. For those with greater experience who are prepared to solder together metal kits, a whole range of possibilities opens up for both locomotives and rolling-stock.

0 gauge light railway _____ 135

The Drewry diesel, BR Class 04 is the ideal choice of small locomotive for a layout such as this. This excellent model is built from a highly-detailed Vulcan kit, but because of the variety of materials used and the detailed instructions, it is quite suitable for the beginner. (Ron Cadman)

In short, a great deal of time will need to be spent constructing the items for this layout as there is so little ready-made stock available for 0 gauge. However, the layout shown, whilst basic, can provide a great deal of detail and is an ideal introduction to the delights of 0 gauge in a comparatively small space.

Further references

Further reading on the Kent and East Sussex Railway:

Branch Lines to Tenterden, Middleton Press

The Kent and East Sussex Railway, S.R. Garrett, Oakwood Press

CHAPTER 9
Railway Preservation Centre

The items which probably interest most modellers are first and foremost the locomotives and rolling-stock. Whether you buy ready-to-run models or build kits, in all probability you will be tempted to acquire attractive locomotives or items of rolling-stock which couldn't possibly run with your other models, such as perhaps a turn-of-the-century design scrapped well before the period the rest of your models are painted to represent.

One answer to satisfying this desire to accommodate a diverse collection of locomotives and stock from different sources, while providing a believable setting, is to make a model of a Railway Preservation Centre.

There are now a considerable number of preserved railways and railway centres scattered throughout the country in both rural and urban environments. These centres differ widely in the type and character of the facilities and services which they offer. Some provide main line runs, others possess only a few hundred yards of track but nearly all usually provide some running track and give rides, and have a collection of locomotives and stock and some form of accommodation for at least part of the collection either in a new purpose-built shed or an old engine shed. Usually there is also some facility for repair and restoration and plenty of dilapidated locomotives and odd items of rolling-stock lying around, either waiting for or undergoing restoration. Finally, there is, of course, car-parking and often a children's play or picnic area and some other attraction such as a model railway or collection of vintage road vehicles.

The plan

The plan (Figure 85) shows the smallest and probably the simplest of the layouts described in this book. It will, at a pinch, provide accommodation for six locomotives, assuming that some will be reasonably-sized prototypes such as 2–6–0 or 4–6–0 tender locomotives. The train would probably comprise of one former main line coach or three or four brake-vans and there is ample storage for these items together with a few others which would represent either spare vehicles or vehicles undergoing or awaiting restoration. Modelling these latter vehicles could be a challenge in itself.

The layout shown occupies an area

Railway Preservation Centre

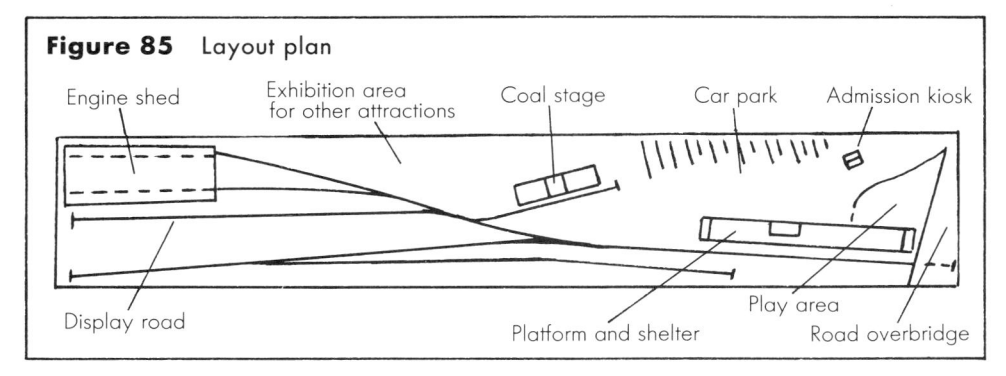

Figure 85 Layout plan

- Engine shed
- Exhibition area for other attractions
- Coal stage
- Car park
- Admission kiosk
- Display road
- Platform and shelter
- Play area
- Road overbridge

of 6ft × 1ft and, in addition to the storage facilities, provides an engine shed and coaling facility and a length of running line. There is also room to provide a rudimentary platform, some additional attractions and car-parking. It is assumed that the centre is on the site of a former steam shed and the overbridge at the right-hand end provides the opportunity for subsequent extension which could perhaps coincide with the provision of a further set of points near the bridge to turn the storage siding into a loop to enable the train engine to run round and return.

Baseboard construction

The layout could be built on a single baseboard or conveniently split into two boards each 3ft × 1ft. Either way, a ½in chipboard surface on a 2in × 1in softwood frame will suffice. Figure 86 shows the construction, and the arrangements for connecting the two baseboards are as described in earlier chapters (brass hinges with removable pins). Pieces of ⅛in plywood 4in tall are fixed to the back and side of one of the boards and will be finished later as a retaining wall. The front is later finished as the road overbridge.

This scene, whilst showing a small shed on the Midland Railway, shows that a realistic shed scene can be created in a small space. The track disappearing beneath the bridge on the left goes to a fiddle yard and enables locomotives to be brought on shed for turning, servicing and stabling and then run off to carry out their duties.

Figure 86 Baseboard construction

- ⅛in plywood backscene
- chipboard surface
- loose pin hinge to join baseboards
- 2in × 1in softwood frame

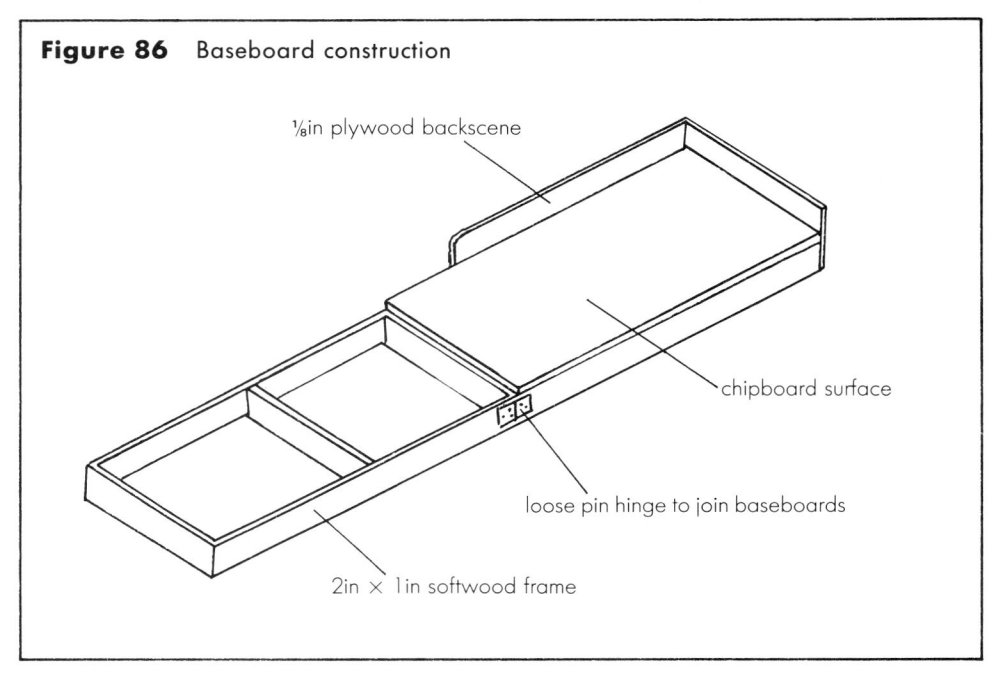

Trackwork

The choice of trackwork for the layout is between the proprietary flexitrack systems such as that of Peco or fine bullhead rail from, for instance, SMP. The former has a wide range of ready-assembled and wired pointwork, while the latter requires you to build your own from kits, but this is not difficult if care is taken and the instructions supplied are followed. The big advantage with the latter type is its finer appearance and, even in this layout, where the trackwork will be largely submerged, its thinner sleepers and finer rail section will make the scenic effect easier to achieve.

The track is pinned to the baseboard and, if the Peco system is used, is ballasted after all is wired and working. The SMP system benefits from being glued to the baseboard, with the ballast being applied whilst the glue is still wet. As always, care needs to be taken to ensure that the track is carefully aligned and flat to the baseboard surface.

Wiring

A wiring diagram showing feeds and suggested section breaks is shown in Figure 87. Point motors should be fix-

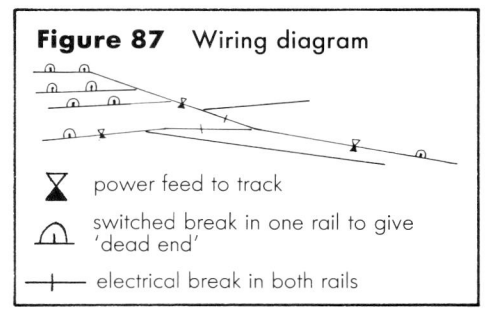

Figure 87 Wiring diagram

- ⏳ power feed to track
- ⌒ switched break in one rail to give 'dead end'
- ─┼─ electrical break in both rails

Railway Preservation Centre

ed beneath the baseboard and, having been wired in accordance with the instructions supplied with them, should be connected to the control panel. This is a demountable box which can be unclipped from the layout when not in use. It contains the controller and section and point switches, the wiring from which is taken in harness via a multi-pin plug and socket to the baseboard. The control panel need only be a simple plywood box and its shape and size will be determined by the type of controller you choose for the layout.

The section switches are important on this layout because they allow one locomotive to be moved off a road whilst another is standing on it. In effect, a number of separate sections are being created along a given track, each fed from the same controller but with a switch between the track and controller allowing any one or combination of sections to be fed with current at any one time. Thus, for example, if you have a locomotive at the end of one of the shed roads and you wish to move it to the coaling stage, having set the road you would need to switch on the section the locomotive is occupying and the one in front of it, assuming no other locomotives are in the way. If the engine you wish to move is in the middle section of the shed road and there is also one at the inner end, you will need to switch off the end section so that the locomotive standing there will not move, and switch on the other one to take the moving engine off to the coaling stage. Careful consideration of the placing of section breaks can increase the versatility of a layout of this nature.

Scenery

If ash-pits are to be incorporated in the layout, now is the time to make them, and they can be simply hollowed out of the baseboard surface between the rails of each shed road in the area at the front of the shed. It is best to lay the track in the ordinary way and then start work on the pits by cutting the sleepers between the rails just inside the rail

This scene shows an ash-pit constructed as described in the text. If you prefer not to make a coaling stage yourself, the Ratio kit, shown here, would be ideal.

Different surface finishes around buildings: while areas of public access would be concreted, traditionally the surfacing may have been of stone sets or cobbles. These could have been repaired and restored at the preservation centre and this is easily represented by the use of the pre-printed embossed cards made by Faller, as shown in this picture.

chairs and removing the necessary length (about 3–6in would be ideal). The baseboard surface is then scored along the edges of what will become the pit; light scoring with a heavy DIY knife repeatedly going over the lines rather than heavy cuts will be easiest and achieve neater results. The area is removed carefully a little at a time with a woodworking chisel, taking care not to go through the baseboard. (Do not despair if you do go through–simply pin a piece of hardboard or plywood beneath the surface.)

The inner surfaces of the pits are now lined with embossed brick card or plastic sheet, and the bottom and lower sides blackened with matt black paint with a hint of rusty brown and blue. Residue left in the bottom of the pit can be represented by a pile of ballast suitably painted.

The ground surface around the shed and track area would in a normal running shed probably be ash. However, as the public will probably be allowed to walk round the shed and its immediate environs, it is quite possible that concrete surfacing would have been laid in these areas. This is easily represented by cutting thick card to fit between the shed roads and this can be extended to provide a walkway between the reception area and the shed, avoiding, of course, the area around the coaling stage. The rest of the surface around the shed, reception area, coaling stage and other sidings would probably be ash, and this can be represented simply by coating the area with a mix of Polyfilla dyed with watercolour powder paint to the desired hue. Later this can be coated with very fine scenic dressings and small depressions could be filled with varnish to represent puddles.

The line on which rides are given would be ballasted in the ordinary way, using either stone chippings or cork granules. Buffer stops should be provided at either end of the running line, and at the ends of the coaling stage road and the two sidings. On a small layout of this nature, it is worthwhile going for the best detail you can and buffer stops are no exception. Mike's Models produce a wide range of types any of which could be used. Add some weeds and debris around them together with the odd oil barrel, spare rail chairs (cast ones are available) and the like.

Similarly, very detailed yet simple-to-assemble water-cranes are available from the Mike's Models range and these should be placed alongside each of the shed roads. They often have a drain below them and this should be set in the layout beneath where the water-bag hangs; it can be streaked with rust, and dabs of gloss varnish will show up as water left on the surface.

The *engine shed* itself can come from a variety of sources, the two most readily available being the Hornby shed with the 'saw-tooth' roof which spans two roads (or three at a pinch) and the Dapol plastic kit. Two of the Hornby and four of the Dapol would provide the required accommodation, but both would need some modification and improvement. The Hornby sheds simply join together, front to back, whereas the Dapol sheds would require the use of two backs, two fronts, four roofs and four walls. Figure 88 shows one of the permutations. The pre-printed Hornby shed can be overlaid with embossed brick plasticard.

Detail the window arches and bays with different brick courses and add stone window-bottoms from strips of 40 thou plasticard painted a grubby

Above *The well-tended garden opposite Haworth station is fairly typical of the broader environment of preserved railways, all part of providing attractive facilities for visitors. An ornamental garden could easily be added around the picnic area suggested for this layout, using small trees intended for N gauge as ornamental bushes, or even those sold for decorating Christmas cakes.*

Below *The water-tower and neatly-painted fence at Haven Street on the Isle of Wight Railway give this scene a model-like appearance. Vehicles parked awaiting restoration can just be seen in the background.*

Hornby produce an engine shed with large side windows which can easily be filled in with embossed brick plasticard. The use of this type of shed is mentioned in the text. Later versions are printed to represent stone but can be overlaid with embossed brick or stone plasticard and their appearance much improved. Separate gutters and downpipes and detailed roof lights with adding glazing bars develop this basic structure into a respectable model, and joining a number together can produce sheds of varying length and width.

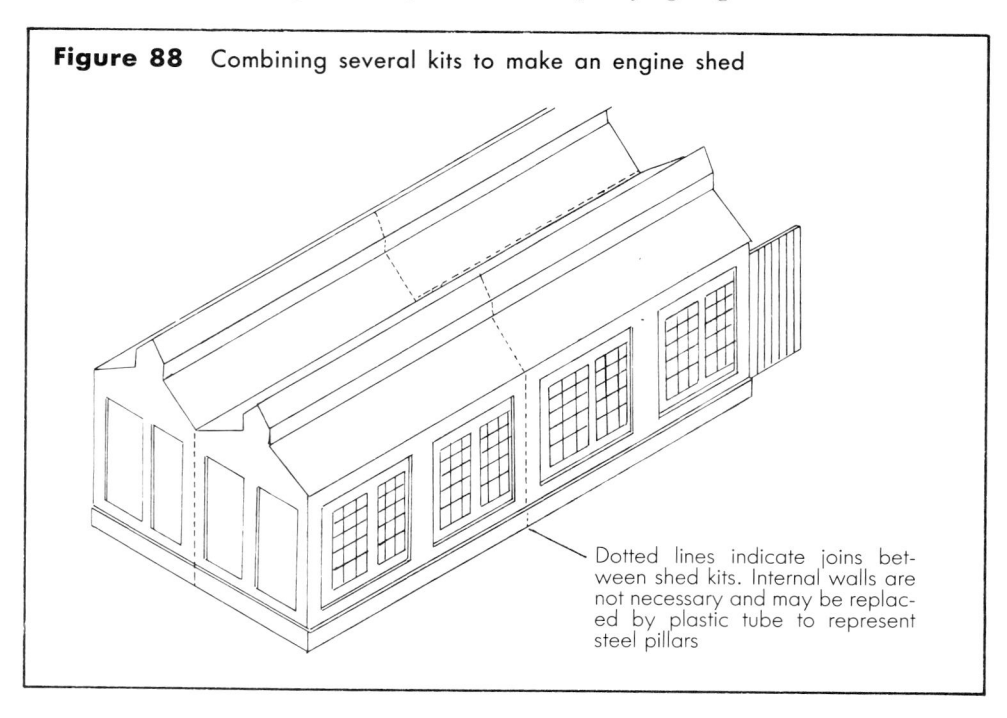

Figure 88 Combining several kits to make an engine shed

Dotted lines indicate joins between shed kits. Internal walls are not necessary and may be replaced by plastic tube to represent steel pillars

Coaling the hard way from improvised facilities is very often the lot of engine crews on preserved lines, as witnessed here at Haven Street. Note the board placed over the bunker side to protect the paintwork whilst coaling is being carried out.

stone colour. The walls should be painted with an overall wash of a matt brick colour, varying the tone slightly by the addition of blues, reds and greys. When this is dry, highlight the mortar courses by flooding the walls with a grey gouache or acrylic paint, wiping it off before it dries. The residue will be left in the embossed courses. Pick out individual bricks in slightly different shades and do the same with the moulded roof tiles. Add guttering and downpipes to the Hornby model using left-over plastic kit parts or rod.

The windows of the Dapol kit will require glazing, while the Hornby shed will need new windows with smaller glass panes and therefore more glazing bars; a surrounding frame should be added using plastic microstrip. Drain gratings can be fixed to the baseboard at the bottom of the downpipes by scraping out a small rectangular hollow and positioning a small grid of microstrip plastic on edge. Don't forget to add the smoke stains above the shed entrance; dry brushing with matt black paint is ideal. Paint the doors and pipes on the shed the matt colour of your choice then tone it down by adding grey. Dry brush a touch of green here and there behind the downpipes and near the gutters to represent discoloration by water. You could, if so minded, detail the inside of the shed, providing roof support columns, brickwork (perhaps whitewashed), work-benches and other paraphernalia. There are quite a few scenic detailing kits which contain tools, barrels and even old boilers!

The *car-park* and approach road would probably be a gravel area, which could be represented by painting the appropriate baseboard area in a 50/50 mix of PVA woodwork glue and water and covering the area in Woodland Scenics fine grade buff-coloured ballast dressing. Use any moulded plastic

Simple model railway layouts

Figure 89 Playground equipment

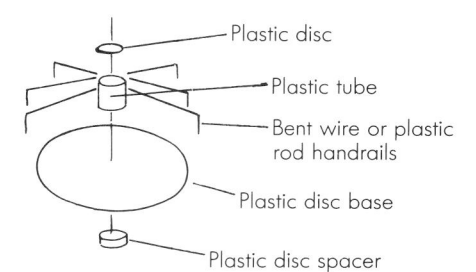

Roundabout
- Plastic disc
- Plastic tube
- Bent wire or plastic rod handrails
- Plastic disc base
- Plastic disc spacer

Climbing frame
- Wire

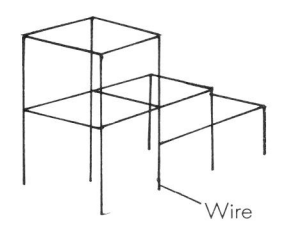

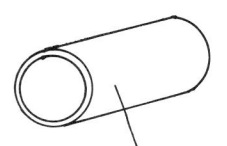

Play tunnel
- Plastic tube painted to represent concrete

Swings
- Cotton 'chain'
- Plasticard swing seats
- Brass rod or tube frame
- 'Bare earth' beneath swings
- Grass tufts at base of frame

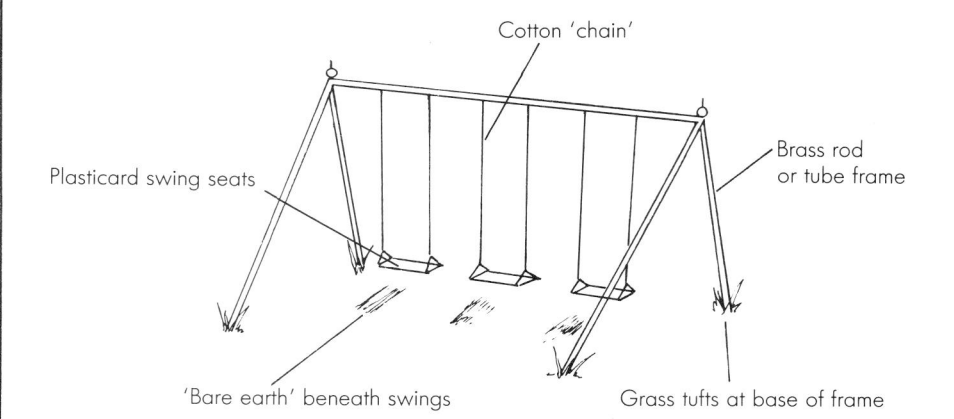

Railway Preservation Centre   145

fence to delineate the car-park area, and provide a wooden hut near the rear of the baseboard as the payment kiosk; working continental lifting barriers could even be installed.

At the extreme right of the layout is a children's *play or picnic area*. It would be a simple matter to make 'adventure playground'-type equipment or even picnic tables and benches, and some ideas are shown in Figure 89.

The *retaining wall* along the back and right-hand side of the layout can now be modelled, and the otherwise flat piece of plywood can be quite easily turned into an effective wall using the suggestions in Figure 90.

The *platform* could be quite a simple affair, representing the type often seen in preservation centres which are built from old railway sleepers, and this par-

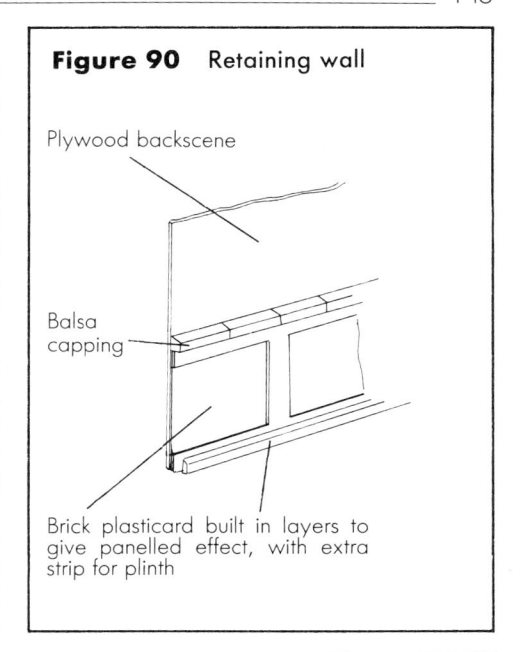

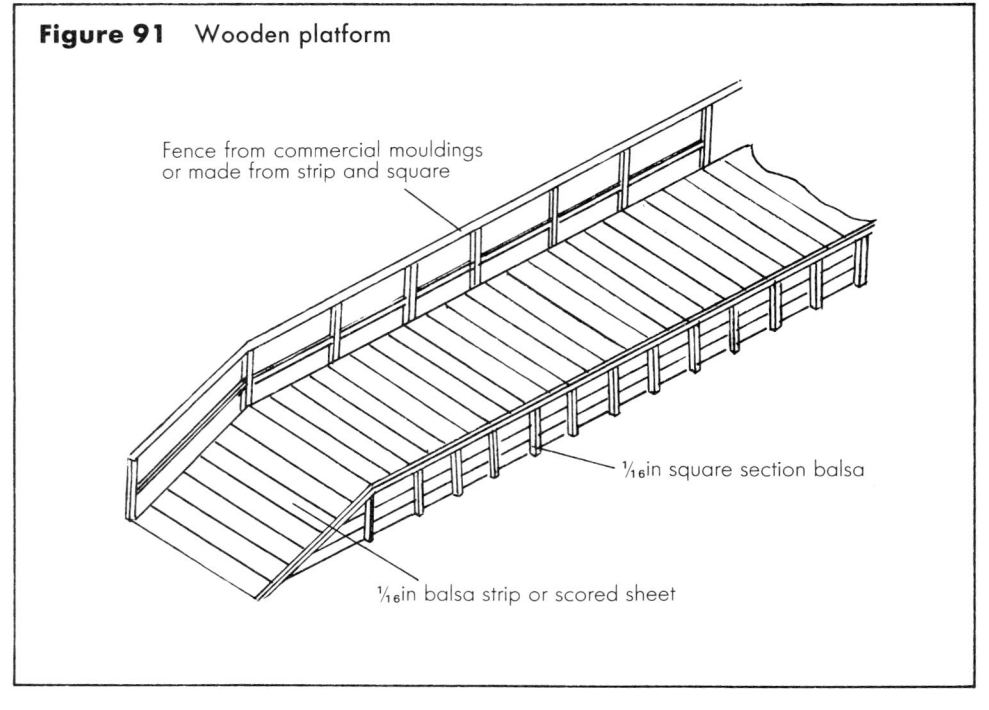

Figure 90   Retaining wall

Plywood backscene

Balsa capping

Brick plasticard built in layers to give panelled effect, with extra strip for plinth

Figure 91   Wooden platform

Fence from commercial mouldings or made from strip and square

$1/16$in square section balsa

$1/16$in balsa strip or scored sheet

ticular platform only need be 12–16in long at the very most. The shorter the length the better to help create the illusion of space. Construction of a simple wooden platform is shown at Figure 91; a simple platform shelter could also be added, easily made from plasticard.

Some provision for a water supply needs to be made. At its simplest, the Dapol water-tower, built as intended, could be installed either immediately to the rear of the shed or further back between the shed and the coaling stage. This kit could be adapted by placing the tank on a simple brick base made from plasticard, or the Hornby water-tower could be used but with the base treated in the same way as the engine shed brickwork.

The right-hand edge of the baseboard is finished as a *road over-bridge* but only half of this need be modelled. One side of a Peco overbridge could be used but, because this feature is at the baseboard edge and thus very likely to get caught and damaged, it needs to be installed around a wooden frame for support and strength.

Remaining scenic features are details and these are largely a matter of personal choice. Obviously, cars and plenty of people spring to mind but, particularly near the shed, bits of locos, wheels, boilers, complete cabs and brake hangers could be added. These can be easily obtained using spare parts and locomotive building components, and by keeping an eye on the junk box at the local model shop. A refreshment kiosk could be adapted from a wooden hut or garage kit, with signs cut from the colour advertisements in magazines, and install dummy point levers at the points.

This layout provides a good excuse for using those relatively cheap kits of ploughing and showmen's engines, vintage road and agricultural vehicles and even some 1/72 scale military vehicles to make a little display which could form another 'attraction' on the site. Some items of rolling-stock can be almost 'scenic' especially those rusty

The usual debris and clutter found around preserved steam sheds is well in evidence here at Haworth on the Keighley and Worth Valley Railway, and the items shown here could well be reproduced on the model to help give the authentic atmosphere of the preservation centre. Spare parts and detailing components could be used to advantage, and maybe a broken or long worn-out locomotive could provide the basis for such details.

Several cheap plastic kits of locomotives and rolling-stock are readily available. Suitably painted, with a cover on the chimney and coupling and connecting rods removed, they soon begin to look like locomotives awaiting or undergoing restoration. Similarly, goods wagons, particularly those obsolete on BR such as cattle wagons, would make good 'exhibits' themselves or, suitably adapted, could also be shown under restoration.

heaps awaiting restoration or partly restored. It could be quite a challenge to re-produce the varied conditions of vehicles seen at preservation centres; a lot can be done with careful painting to represent dereliction, and the photograph shows a fairly basic attempt at this. Some items of stock are merely sheeted over, and well-painted aluminium foil with cotton 'rope' can be used to represent tarpaulin sheets.

Locomotives and rolling-stock

There is not really a lot of operational scope for this layout, but some interesting operations could be gained by moving locomotive and stock around for coaling and watering. The layout does, however, provide a realistic home for *Flying Scotsman*, *Stowe* or *Coronation* in a mere 6ft × 1ft! Whilst appreciating that there is often little desire or indeed need to improve the latest commercial models, some detailing is usually possible, even if it is only some real coal in the bunker or tender and the painting of silver handrails, wheel rims and valve gear.

This layout can provide a useful step to other areas of modelling because it can legitimately house anything which takes your fancy; if you try your hand at building a locomotive or wagon kit and get hooked, you will have a ready-made home for your latest creation. Also, because it is a small layout, you can go on adding details and upgrading it regularly as your skills and interests develop.

CHAPTER 10
N gauge North American layout

Last, but by no means least, a suggestion is offered for a layout which is intended purely as a 'fun' layout without too much regard for correct practice or prototypical operation. It provides for a fair amount of scenic development and an opportunity to incorporate a variety of bridgework, including, perhaps, one of those spindly wooden affairs so often the location of cinema melodrama! The layout also provides an opportunity to utilise some of the colourful North American models, so many of which are readily available at very reasonable prices in this country.

The plan

The layout (Figure 92) occupies an area

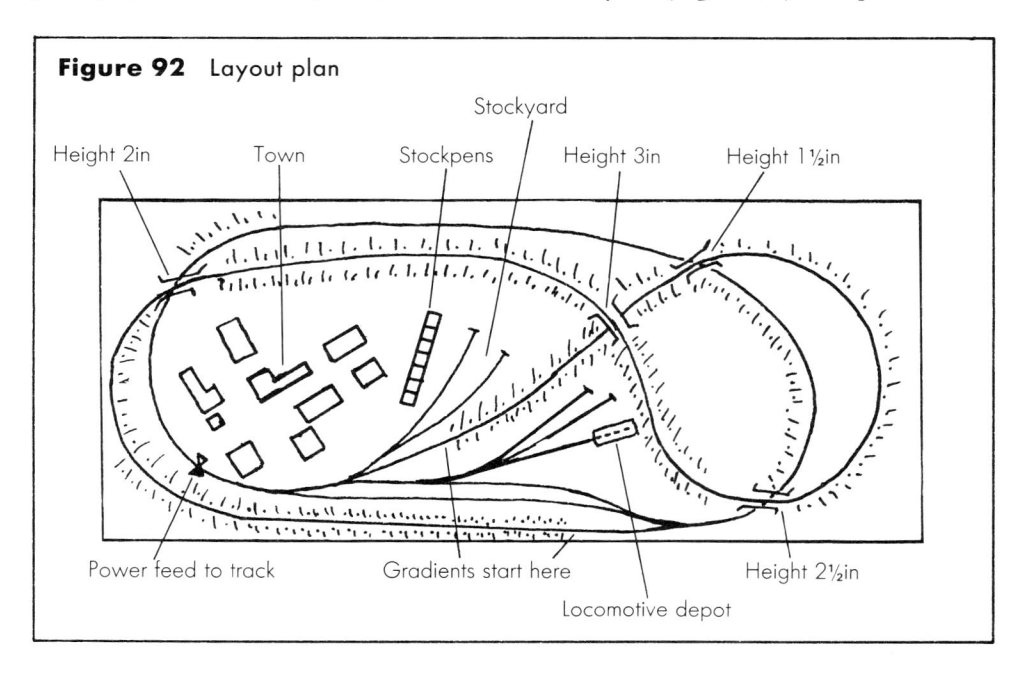

Figure 92 Layout plan

N gauge North American layout

of just 5ft × 2ft and provides a single track system of one level circuit and a loop which climbs and twists above and around the main circuit. It is this loop which requires the bridgework to carry it over the lower circuit. The facilities shown are suggestions only, but since the layout gives the impression of being a junction it is possible to justify some sidings and a locomotive establishment including not only the provision of water and fuel but also the facility to change locomotives. It may be that you will prefer to give more emphasis to one of these suggestions at the expense of the other.

Baseboard construction

Construction starts as usual with the baseboards. There are several methods of building baseboards for multi-level layouts of this nature, and they basically fall into two categories, variations on the 'open top' or 'grid' system and building up from a solid base. With the 'open top' method it is essential that the layout is worked out full size, on, say, decorators' lining paper, the final plan being the template for actual construction of the baseboards and, in particular, the location of the trackbed supports. The method starts with an open framework the size of the finished layout, then softwood 'risers' or supports are added to the basic frame to support the trackbed which is cut from plywood.

This open framework has two main advantages. Firstly, it is strong and rigid but easily allows the trackbed to rise or drop below the basic level of the frame. Secondly, it is easily adapted to provide support for additional trackwork which can be added without the restrictions sometimes encountered when adapting or extending

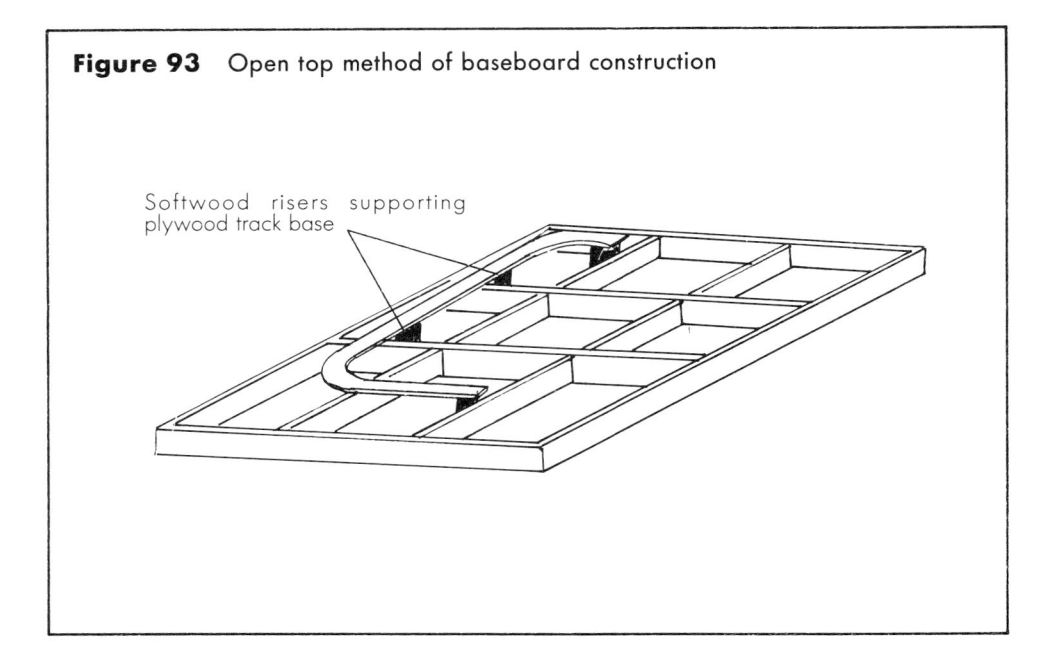

Figure 93 Open top method of baseboard construction

a layout on a solid-topped baseboard. This method of construction also has the advantage of enabling quite dramatic scenic development to take place both above and below track level. Figure 93 shows the principle of this type of construction.

The other more traditional method of baseboard construction uses the conventional softwood frame covered with a Sundeala or chipboard sheet. The basic ground level circuit is marked on this and the higher trackwork is carried on plywood bases, about 1in wide at most, cut to the curvature of the track. These are supported on softwood blocks, carefully cut to height to ensure a smooth gradient for the trackwork. The blocks are glued and pinned to the flat top of the baseboard, and Figures 94 and 95 show this method of baseboard construction.

Either method will require careful planning and if you intend using commercial bridge kits, due regard will need to be paid to their height.

The plan (Figure 92) suggests heights for the various levels of trackwork; these are for guidance only, but the higher level track must be raised at least enough to provide clearance for trains passing beneath. If you decide to lift the track much higher, bear in mind that the combination of a steep gradient and tight curves will probably severely limit the length of trains which can be hauled. Finally, remember that whatever method of baseboard construction you use, start from the bottom and work up.

Trackwork

Moving on from the baseboards to the trackwork, flexible track and ready-made pointwork such as the Peco Streamline system would be suitable for this layout. It could be laid on the moulded foam underlay which is designed to be used with it, and this should help to quieten the running, particularly on the thinner plywood

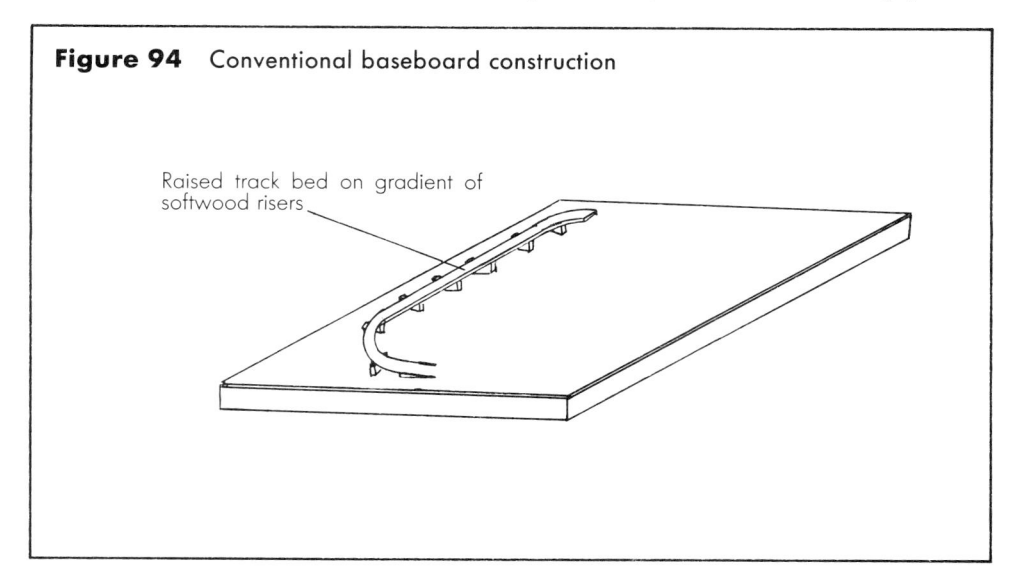

Figure 94 Conventional baseboard construction

Raised track bed on gradient of softwood risers

N gauge North American layout   151

Figure 95   Gradient formation

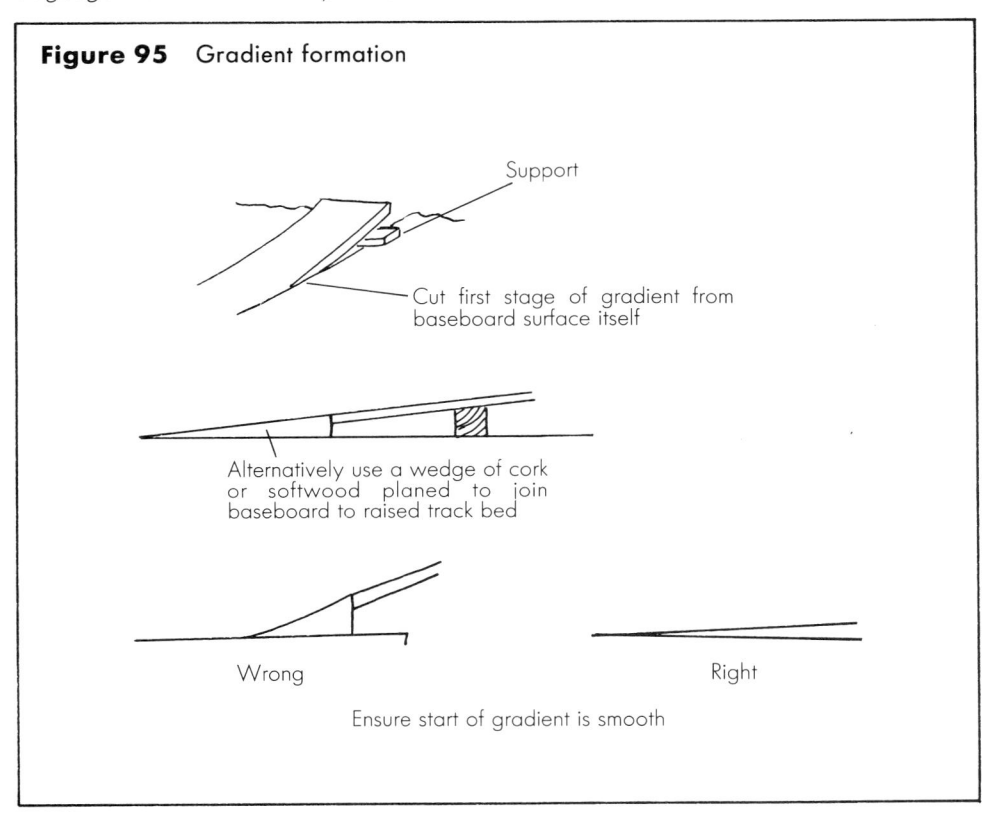

Support

Cut first stage of gradient from baseboard surface itself

Alternatively use a wedge of cork or softwood planed to join baseboard to raised track bed

Wrong   Right

Ensure start of gradient is smooth

bases used on the high-level trackwork. Alternatively the trackwork could be pinned directly to the baseboard and ballasted afterwards with coarse sand over which a 50/50 mix of PVA glue and water is sprinkled with an eye-dropper or similar. A combination of the two, that is laying the track on the foam underlay and then adding a thin dusting of sand ballast glued in the same way as in the sand-only method, would provide a useful compromise between sound insulation and appearance. Do, however, avoid gluing the point tie bars when adding the ballast.

There is a lot of curved trackwork on this layout and this is best laid with the aid of templates which will ensure smooth curves and, of course, aid running.

Wiring

The track now needs to be tested and wired. You could get away with using only one feed to the layout, but versatility can be increased with additional wiring to enable, for example, a train to circuit whilst shunting takes place. This would, however, require the use of two controllers. The plan (Figure 92) shows the position for a single feed, but the final wiring will depend on

Above *A section of layout under construction using wire netting on a framework of scrap softwood. The hardboard in this case will be used for a small section of trackbed but could also form a road or other feature.*

Below *The application of squares of plaster bandage to the wire netting frame in progress.*

Bottom *The view from behind the scenery.*

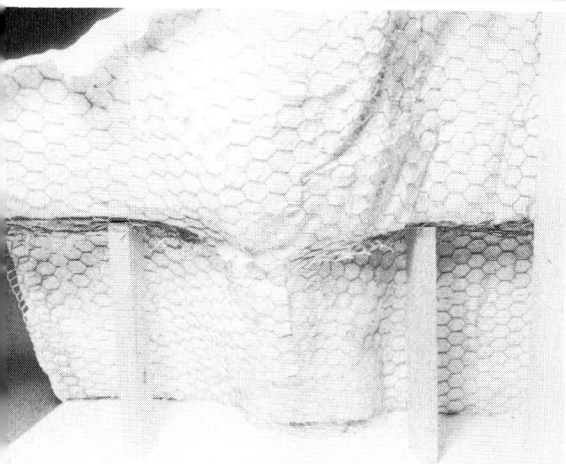

how you adapt the plan and on the facilities you incorporate.

As the layout is constructed on a single baseboard, there is no need to split the wiring, so it can be run straight back to the control panel. The panel could be neatly and unobtrusively installed within the layout if panel-mounting controllers were used. The layout should only have the reduced voltage required to run it connected permanently, so for safety, if a separate controller transformer system is used, the transformer should be placed in a separate box beneath the layout. As always, follow the manufacturer's instructions when installing your control equipment and if you have any doubts consult a qualified electrician. An electronic controller, which compensates for gradients and tight curves to enable trains to maintain a constant running speed, would be ideal for this layout.

Scenery

The first task on the scenic side is to paint the trackwork with a dilute mix of track colour and pick out the rail sides with rust. This is a tedious job but if done carefully is well worth it.

The basic ground cover is sand, or sand/earth colour scatter material, which is glued to the baseboard with a 50/50 mix of PVA glue and water. This can be detailed later with the addition of boulders, vegetation, bushes and clumps of weeds. The rest of the scenic treatment basically involves providing the 'mountains' through which the railway climbs. These are built on a framework of softwood covered in wire netting which is shaped and contoured to suit the effect you wish to create. Once the basic framework has been established, it can be covered with plaster bandage or papier mâché. Cover the trackwork with tape to avoid covering it with plaster and paint during this and the next few operations.

There are two basic methods of representing the bare rock surfaces and outcrops, and a combination of both would probably provide the best results. The first method is to coat the base with a stiff mix of Polyfilla which is sculpted before it is dry to achieve the desired effect. To get a really authentic effect, reference to photographs in geography or American travel books will help.

There are in turn two variations on the second method, and both involve very basic moulding. The first variation involves using a mould of aluminium kitchen foil. A sheet of foil is crumpled up then opened out and spread over the wet plaster surface. When the plaster has hardened the foil is removed to reveal a quite effective rock finish.

The second variation involves the moulding of individual rocks. Select a suitable specimen which has features similar to the effect desired on the model; large pieces of coal are often suitable. The mould itself is made from the rubber solution sold by art and craft shops for making moulds for plaster of Paris garden gnomes and the like. Thoroughly wash the 'master' rock, then make the mould by coating it with several coats of the rubber solution. When cured, the rubber mould is removed and the rocks for the layout can be cast by pouring a thick mix of plaster into the mould which, when the plaster is dry, can be peeled off and re-used. The whole process can be speeded up and variety ensured if several moulds are made and used simultaneously. The resulting rocks can be fixed to the scenic base with glue, or preferably bedded into the wet plaster surface. Any unsightly gaps or cracks could be filled with plaster.

Colouring and detailing the basic scenery is important to create the right effect. The main thing is to avoid strong, bright colours which can spoil the look of an otherwise excellent layout. Build up the colour gradually

Detailing the scenery: the plaster bandage covering has been applied in overlapping 6in or 8in squares which can be applied folded and curled up to give the impression of rock strata, but further detail work will be required to add greater shape and character to the rock formations. This is achieved by adding coats of Polyfilla which is sculpted with a knife when almost dry; this process has been started towards the top left-hand corner of the picture. A 'rock fall' has also been started at the bottom of the picture using Polyfilla powder applied to a dampened surface; small rocks, coarse sand etc can also be added to give the desired effect. The ground immediately below the rock face is a roadway, so the addition of a sign warning against rock falls would add authenticity. Add some vegetation in crevices and on the flat surfaces of outcrops using sparsely-applied scenic scatters of suitable colour and tufts of foliage mat.

154   Simple model railway layouts

Above *There is ample opportunity in this layout for some exciting bridgework. As can be seen here, the construction of bridges is a straightforward process using commercial plastic bridge sides of various types, while the track itself is carried across the gap on a hardboard base.*

Below *The finished mountain ledge. The track has been laid in the usual way and ballasted with sand.*

with thin washes of watercolours. Similarly, the vegetation should not be bright green scatter material, nor should there be even colouring; subdued colours should be used and the scatter material mixed and blended from different colour packs to give the desired effect. For this layout, vegetation is likely to be scrub with the odd patches of grass at the bottom of slopes, in crevices and on flat areas amongst the rocks. The bushes and scrub can be represented by tufts of Woodland Scenics foliage and sprigs of lichen.

An alternative to the use of scatter material which would be very effective on this layout is a technique called 'zip-texturing'. Zip-texturing is common in the USA and basically involves sieving a mix of plaster of Paris powder and watercolour powder on to a damp plaster surface. The effect is quite a fine texture in which the colour is subdued by being mixed with the plaster. It is worth experimenting with this technique as some very interesting effects can be created, and subtle changes in colour become quite easy.

The last detail which can be added to the rock is the illusion of a rock fall. Zip-texturing could provide the finer build-up of deposit, but crushed rock (lumps of plaster broken when dry) in layers, the biggest and heaviest at the bottom, can be quite effective and add some character.

A number of scenic ideas for the trackwork as it climbs and twists through the mountains could be considered. One is the provision of a bridge section where the track bed clinging to the mountain side has been washed away. Snow sheds or simple wooden shields to protect the line

N gauge North American layout

from rock falls are also easily added.

Some bridgework will be required to carry the lines over one another. There are plenty of plastic kits available for all types of bridges, from simple iron girder types to spindly wooden trestles; the choice is yours. Whatever you choose, the end result will be improved if the plastic is painted and weathered with matt browns and greys on wood, and rust streaks on metal bridge sections, after the main colour has been added.

The buildings and railway structures on the layout will depend on the final choice of track plan, sidings, locomotive facilities and so on. There is a wide choice of plastic building kits to cover engine sheds, railway depots, town buildings and industrial structures. Stockyard pens can easily be represented using plastic fencing and adapting this with plastic strip to provide gates. Don't forget to add plenty of cattle which would benefit from a wash of pale dusty hues and may even require a repaint as the majority seem to be the black and white Friesian type rather than the yellow or brown-coloured breeds common in the American mid-west.

Whatever buildings and structures you choose, they will benefit from some work with the paintbrush to remove the glossy plastic finish. The use of washes of well-thinned greys, pale browns and ochres will tone down the colour and help them to blend into the layout. Road vehicles should be similarly treated, and here again there is quite a choice for this layout, particularly if you are prepared to seek out retailers who specialise in American railroads. You will be surprised how many there are in the UK, and a glance at the model railway magazines listed at the end of Chapter 5 will reveal quite a number.

If locomotive facilities are provided, they will vary depending on whether steam or diesel locomotives are used. Most modellers will probably run both

The old-time 4-4-0 is a Bachmann model, and the coaches are by Arnold. Bachmann also produce old-time coaches and freight stock, thus enabling a 'Wild West' period to be recreated in N gauge.

Modern diesels are available from a number of manufacturers and are quite colourful in their American liveries. Here, a Bachmann U-36B in the Santa Fe livery of silver and red is shown. Two types of bridge can be made from this type of girder section; by the simple expedient of fastening them upside down, an underslung bridge can be created.

and if this is the case the depot should be equipped for steam with a storage tank and facilities for refuelling diesels which, if a suitable plastic structure is not available, could easily be built from scrap materials.

Locomotives and rolling-stock

There is a vast range of N gauge equipment available, and whilst this layout is not intended as a serious representation of a particular system or period, it would be worth bearing in mind that modern bulk capacity stock should be hauled by diesels. An 'old-timer' 4-4-0 would look a bit out of place with a train of modern bulk grain wagons and conversely an SD40 diesel would not look right with an old-time combine coach!

Further references

All About N Gauge Model Railroading, Paul Garrison, TAB Books

Model Railroading in Small Spaces, Paul Garrison, TAB Books

Index

Allan, Ian 32
Ash 12, 14, 21, 31, 108, 125, 127, 140
Ash-pit 139

Backscene 29, 36, 42, 75, 78, 109
Ballast 12, 13, 14, 27, 44, 45, 59, 76, 77, 83, 84, 95, 108, 124, 140, 145, 151
Balsa wood 12, 46, 68, 83, 86, 113, 131
Barrels 20, 52
Baseboard construction 10, 11, 12, 25, 26, 42, 43, 57, 74, 90, 122, 137, 148, 150
Baseboard joints 10, 26, 44, 76
Bicycles 20, 31, 52
Bodiam 125
Brass rod 12, 26, 75
Brick paper 52
Bridges 27, 28, 46, 48, 49, 155
Bristles 20, 31
British Railways 24, 36, 41, 49, 71, 104
Buffer stops 93, 99
Bullhead rail 12, 26, 43, 93, 138
Bushes 65, 86, 99, 109, 130

Camomile tea-bags used as grass 84
Cardboard 16, 32
Car park 143, 145
Cattle facilities 24, 34, 78, 79, 83, 84, 127, 155
CCW 134
Chicken wire 63
Chipboard 74, 76, 90, 97, 122, 137, 150
Chisel 29, 140
Church 34

Coal dust 14
Coal facilities 24, 40, 41, 51, 139, 140
Control panel 15, 28, 45, 51, 59, 60, 77, 97, 109, 123, 139, 151, 152
Copper electrical tape 59
Cork ballast 14, 124, 141
Cork sheet underlay 27, 76, 123
Craft knife 77, 97, 140
Crane 18
Cranks 36
Crates 36

Dairy 122
Dapol 34, 51, 141, 143, 146
Dart Castings 31, 84
Das modelling clay 46, 97, 129
Door bolts 28
Dry brushing 51, 52, 80, 86, 143
D & S Models 38
'Dunster Station' 32
Dustbins 36

Eggerbahn 101
Electrical joint 10
Embankment 21
Embossed stone card 28, 32, 46, 140
Emulsion paint 31, 85
Engine shed 14, 16, 20, 21, 139, 141
Eye-dropper, use of 95, 108, 124, 151

Faller 32, 68
Farish, Graham 67, 69, 71
Feeds, electrical 15, 27, 28, 59, 77, 78, 108, 123, 139, 151

Fencing 34, 63, 69, 127, 131
Fiddle yards 25, 26, 28, 41, 42, 43, 48, 58, 74, 80, 90, 113
Fine-scale track systems 12
Fishermen 20, 31
Fish-plates 13, 14, 77, 93, 108
Flat-bottom rail 26, 43, 44, 109
Flexitrack 12, 26, 44, 57, 76, 93, 95, 108, 123, 138, 150
Foam rubber 123
Foam underlay 57, 58, 59, 150, 151
Foliage 20, 31, 86, 109, 128
Foliage mat 99, 128, 129

GEM 85
Gibson, Alan 26, 122
Glass-paper 113, 125
Goods shed 24, 34
Goods warehouse 16, 20
Goods yard 51
Grass 31, 32, 84, 128
GWR 23, 24, 32, 34, 36, 37, 41

Hardboard 29, 140
Heljan 52, 68
'Highworth' 122
Hinge pin 12, 75
Hinges 12, 26, 75, 107, 109, 137
Hornby 32, 48, 52, 141, 143, 146
Horse-drawn carriage 16
Jigsaw 10
Jouef 111, 113

Kadee 74
Kent & East Sussex Railway 9, 122, 125
Keyser, N. & K.C. 117
Kitchen foil 153

Langley Miniature Models 64, 85
Layout plans 10, 23, 39, 41, 57, 72, 90, 102, 121, 136, 148
Leleux Index of Drawings 32
Letraset 68
Lima 36, 52
Lining paper 13, 149
Livestock 31, 36, 63, 64, 80, 99, 106, 131
LMS 24
'Lochtay' 121

Locomotives 16, 21, 22, 23, 24, 36, 37, 40, 41, 42, 52, 55, 69, 71, 72, 86, 101, 103, 104, 115, 116, 118, 119, 122, 131, 132, 133, 134, 135, 147, 156
Logs 18
Loop 10, 24, 26, 28, 96, 123

Masking tape 27
Melamine strip 26
Merten 113
Mikes Models 18, 141
Milliput 80
Minitrix 68
MKD 111, 113, 115
M & L Models 38
Moulded chairs 13, 77
Multi-pin socket and plug 15, 16, 28, 45, 46, 51, 77, 109, 124, 139

Newspaper 63
Northiam 125

Oil terminal 78, 83

Packages 36
Painting 16, 30, 62, 64, 78, 80, 84, 85, 111, 115, 125, 140, 143, 154, 155
Papier mâché 63, 153
Paths 21
Peco 32, 43, 44, 46, 51, 57, 65, 69, 71, 76, 93, 108, 122, 123, 124, 125, 138, 146, 150
Piano wire 60
Plaster bandage 63, 97, 153
Plaster of Paris 153, 154
Plasticard 28, 32, 34, 51, 67, 69, 127, 141, 146
Plastic kits 18, 34, 67, 68, 86
Platforms 16, 18, 32, 34, 51, 97, 125, 145
Playground 34, 36
Pliers 63
Plywood 10, 11, 25, 26, 28, 42, 43, 62, 74, 75, 90, 92, 97, 124, 127, 137, 139, 140, 145, 149, 150
Point blades 14, 76, 84, 93, 95, 108
Point frogs 44, 57, 108

Index

Point kits   12, 15, 26
Point levers   93
Point motors   13, 15, 26, 28, 59, 60, 93, 97, 99, 108, 124, 138
Point rodding   36, 131
Point tie bars   27, 95, 123, 151
Pointwork   13, 25, 27, 43, 44, 57, 58, 76, 93, 123
Pola   68
Polyfilla   31, 64, 85, 95, 97, 99, 109, 111, 127, 140, 153
Polystyrene   78, 99, 109, 127
Pond   20
Preiser   20, 65, 113
PVA woodworking glue   13, 27, 58, 84, 95, 99, 108, 123, 124, 128, 143, 151, 152

Rail chairs   140
Rail joints   10, 14, 27, 45
Ratio   34, 36, 38
Razor saw   76
Reeds,   20, 31
River   28, 29
Road overbridges   16, 21, 42, 46, 48, 62, 113, 124, 137, 146
Road vehicles   16, 51
Rocks   20, 30, 31, 63, 97, 99, 153, 154
Roco   101, 118
Rolling-stock   13, 14, 21, 22, 24, 25, 36, 37, 40, 41, 52, 57, 69, 71, 72, 86, 101, 115, 117, 118, 119, 122, 131, 132, 133, 134, 135, 147, 156
Romney, Hythe & Dymchurch Railway   9
Rubber solution   153

Sand   30, 31
Sawdust, dyed   65
Scale Link   20, 31, 65, 68, 128, 129, 130
Scenic details   16, 36, 52, 60, 64, 65, 67, 68, 78, 80, 85, 86, 99, 131, 146
Scenic materials   14, 20, 63
Scenic scatter   14, 16, 20, 21, 65, 95, 128, 152, 154
School   34, 36
Screws   75, 76
Sector plate fiddle yard   25, 26, 28, 43
Shinohara   76

Sidings   25, 26, 90
Signals   36, 46, 131
Signal box   24, 34, 51, 60
Sky paper   36, 78, 109
Sleepers   51, 76, 93, 99, 108, 123, 139, 145
SMP Scaleway   12, 26, 138
SNCF   104, 105, 115, 116, 117
Softwood   29, 74, 76, 90, 122, 149, 150, 152
Softwood frames   10, 75, 92, 137
Softwood strip   18
Springside Models   85, 134
Stations   16, 34, 60, 68, 113, 127
Station buildings   25, 32, 42, 48, 49, 68, 112, 125, 155
Steps   49
Stephens, Colonel   16, 115, 125
Streams   25, 27, 28, 30, 34
Street lamps   136
Stud and probe   15, 97
Sundeala   10, 11, 25, 26, 42, 107, 150
Surgical lint   128
Switches   15, 16, 44, 123
  DPDT   96, 108, 139

Telephone kiosk   36
Template   93, 123, 149, 151
Trackbed   25, 149
Track pins   58
Track plan   13, 24
Tractor   64
Transformer   77, 152
Trees   21, 64, 109, 129, 136
Tunnels   62, 63
Turnouts   12, 15, 26, 43
Twigs   18, 32, 131

Varnish   20, 31, 64, 78, 141
Vulcan   134

Warehouse   40, 52, 73, 78, 80, 83, 84
Water crane   141
Waverley   123
Weeds   14, 95, 124, 152
Wet-and-dry paper   65, 68, 77, 125
Wills scenic range   28, 112

Wire, multicoloured   109
Wire netting   97, 152
Wire wool   31
Wiring   14, 15, 27, 77, 95, 108, 123, 124, 138, 151

Woodland Scenics   84, 86, 108, 112, 128, 129, 143, 145, 154
Wood-yard   18

Zip-texturing   154